DEMOCRATIC CONTRACT

(FOR ALL TYPES OF CONTRACTS)

KANHAIYA JEE ANAND

and

Team of Engineer Live Foundation

Title : Democratic Contract
Author : Kanhaiya Jee Anand & Team of Engineer Live Foundation
Edition : 1st (January, 2024)
ISBN : 9788196784676

Published by

A Venture by -
PRACHI DIGITAL PUBLICATION

Regd. Add.: 254, Khuriyakhatta No. 10, Bindukhatta,
Lalkuan, Nainital - 262402, Uttarakhand, India
Website : www.taneeshapublishers.in
E-mail : taneeshapublishers@gmail.com
Phone : +91 845481 2712, +91 976041 7980

Printed by :
Manipal Technologies Limited, Bengaluru - 560001, Karnataka

Acknowledgements

ENGINEER LIVE FOUNDATION (ELF) extends special thanks to the following members of its update Task Group: Sri Kanhaiya Jee Anand Director ELF, Sri N. K. Suman Director ELF, Sri Raman Kumar Rajesh Director ELF, Sri Kumar Vimal, Director ELF, Sri K K Kalyan, Director ELF Sri Arun Kumar DGM BSRDCL, Sri Rajnish Misra, Director ELF, Sri P. K. Srivastava, Founding Member ELF, Sri M. Madhusudan, Member ELF, SriKK Shahi, Project Directors Shinghla, Shri Pradyot Kumar, Vice President GRIL, Sri M.S,Rawat,Director Aecom, Sri Manoj Kumar, Director ELF & Project Manager, C&C, Sri K. Parameswaran Project Manager, NECL,Smt.Sharda Kumari, Director ELF, Sri Kalyan Ji, Director ELF, Sri Ravi Shankar, Director ELF, Sri OM Prakash, Director ELF, Sri Nitesh Kumar, Executive Director L&T, Sri Manoranjan Kumar, Director ELF, Sri Manish Kumar, Tata Sons, Smt. Nirija GM Railway, Sri Mukesh Kumar, Ex- Director Railway, Sri Ashutosh Sinha, Ex Joint Secretary PMO,Sri B.K.Jha Vice President Rodic, Sri Prabhakar Kumar, ELF Rajeev Maini, Director, CEAI

The preparation was carried out under the general direction of the ELF Contracts Committee which comprised Kanhaiya Jee Anand, Director ELF, Sri Krishna Kumar Retired Judge, Jharkhand Govt.Ms. Bedatri Anand, Architect ELF & Sri Utakars Anand ArchitectELF.

Drafts were reviewed by many persons and organizations, including those listed below. Their comments were duly studied by the Update Task Group and, where considered appropriate, have influenced the wording of the clauses. Sri Krishna Kumar, Judge Jharkhand Government, Sri N.K. Suman Director ELF, Sri Sandeep Kumar Member ELF, Sri Sandeep

Kumar, Member ELF, Sri Pawan Vutukuru, Member ELF. Acknowledgement of reviewers does not mean that such persons or organizations approve of the wording of all clauses.

ELF INFRASTRUCTURE DEVELOPMENT & DEMOCRATIC CONTRACT wishes to record its appreciation of the time and effort devoted by all of the above.

The ultimate decision on the form and content of the document rests with ELF DEMOCRATIC INFRASTRUCTURE DEVEL-OPMENT & MAINTENANCE WORK.

Foreword

ENGINEER LIVE FOUNDATION (ELF) published, in 2023, First Editions of One new standard forms of contract:

BASIS of Contract is Implementation in right approach such as Extreme Quality of final product should be achieved, such as LifeSpan of final product should be at a higher level.

Development of society needs relations. Initially relationshipdeveloped in between the mother and Child, further Man and Women for population growth. To safeguard the relationship, People started to develop their Living House, Road, Nalla, Well, Dam, School, College, Office, Social health center etc. by collectiverelation to each other. Even for agriculture works, people are helping each other to do the agriculture work. Later and similarly this relation developed in the form of Contract too, due to economic development. Hence in the world, from the beginning of development process, following relation is in practice. Same relation now has developed in the form of Contract. Followingtypes of Contract Document is in force in the world today:

1. Cost Plus Contract
2. Lump Sum Contract or Engineering Procurement &Construction (EPC) Contract
3. Public Private Partnership (PPP) Contract
4. Family Contract - In Islamic Religion Marriage is also treatedas a Contract, while this is the initial relation developed between Man and Women
5. Democratic Contract - Every Individual is equal in all terms, if opportunity is available, everyone has equal opportunity to Live,

equal opportunity to work

ENGINEER LIVE FOUNDATION taken consideration of Democratic View where Govt is by the People and for the people, hence in Democratic Point of view every people in Concerned Country have the equal Authority, hence Contract should like Democratic. Democratic Contract accepting all types of contractsas desired and required.

Two types of Contracts are considered for the Democratic Contract is as under-

1. INFRASTRUCTURE DEVELOPMET & MAINTENANCE WORK AND
2. EMERGENCY WORK

   I. INFRASTRUCTURE DEVELOPMENT & MAINTENANCE WORK is prepared for all typesof Infrastructure Development Work in a Country.ENGINEER is the key resource person for this type of CONTRACT.

   II. EMERGENCY WORK CONTRACT is prepared for Emergency of Work. This is for shortterm basis contract normally In Charge of system,organization is responsible to run this Contract.

In Development of Infrastructure, Maintenance Work and Emergency Work following Team is essential to derive is as under:

EMPLOYER - Responsible to appoint the Authority and Architect & Engineer as per finalized any Developmental Work. Employer deploying both teams. President is called the Employer for India Government and Governor is called the Employer of the State Government.

For Example, Suppose Bihar Govt has finalized to construct the Ganga Development Work. Then the Governor will appoint the

Authority and Engineer for that work.

Authority - The Authority is a body or a corporation to administera revenue producing public enterprise.

For example, for the Ganga Development Work Project, the Authority shall be responsible for all liaising as well as arranging the Land, Utility Services and Arrangement of Funds.

Architect & Engineer - Architect & Engineers, as practitioners of engineering, are professionals who conceptualize, design, analyses, build and test machines, complex systems, structures, gadgets and materials to fulfil functional objectives and requirements while considering the limitations imposed by practicality, regulation, safety and cost.

Architect & Engineer must first divide the work in various parts or stages and similarly as per R&D, all required design should be finalized for maximum life span in years and by bidding process,the said project work should be awarded to experienced, competent and qualified Contractor. Appointment of Architect & Engineer should be done for until service life span of the project.

Contractor - Contractor should be selected separately for each ofthe divided/sub-divided work and should be appointed to constructas well as take care of constructed works till the End of service lifeof the said work.

Works of Various Party

AUTHORITY - Responsible to Organize Land Acquisition, Utility Shifting and Procurement and Management of Funds. Attending Monthly meetings should be organized by ENGINEERto understand the position of Land Acquisition and procurement/planning of Funds. Attending Tri-Monthly Meeting organized by ENGINEER with CONTRACTOR to understand and Monitor the Progress of various Works.

ARCHITECT & ENGINEER - Responsible for Conceptualization, Research, Inception & Design for planned Civil, Mechanical and Electrical works, Bidding, Finalization of Contractor, Monitoring of Procurement and Checking of Quality, Billing & Recommendation of billing for Payment to the Authority, Regular Planning for Maintenance, Monitoring, Checking & Billing of Maintenance Work. Organizing of Weekly Meeting with Contractor and apprising, recommendation to Authority pertaining to Availability of Land, Fund requirement and Progress of Workon monthly Basis.

Their responsibility of teaching to the Engineering Student in Engineering College with practical knowledge. Either various Engineering college to be under control or new college also be established by Architect & Engineer to develop the Engineer with present knowledge for the future.

CONTRACTOR - Responsible to Execute the Work in Time as desired by Authority/Client. Contractor will organize the Daily inhouse Meeting with their own Employee.

Employee of Authority, Architect & Engineer And Contractor: After completing/working for the minimum period of 5 years with an Organization, Employee can be allowed to move to Authority or to Architect & Engineer and or Contractor for securing the position as per their expertise as well as their interest of work. Similarly, Architect & Engineer can approach to Authority and Contractor and in the same pattern an employee can move from Contractor to Authority and or Architect & Engineer, as per their expertise & interest of Work and availability of vacancy.

For single Work, the same AUTHORITY and Same ARCHITECT & ENGINEER should continue for similarity and unique Research and Design. For Example, let us Suppose Government sometime decided to develop a GANGA DEVELOPMENT PROJECT for Complete reach/length/stretch of GANGA. Government must select the AUTHORITY and ARCHITECT & ENGINEER TEAM.

WORK - GANGA DEVELOPMENT PROJECT CORPORATION (2520) km and 5 km of both side of Ganga) Office of Authority and Engineer should be developed to handle the Work. All strategic planning includes Team of Dispute Redressal, Fund Management Model from each Engineer Representative should be discussed and finalized for the future.

AUTHORITY- Selection should Government Firm or Firm which is under control of Government such that coordination should become easy. Selected Firm should continue for Life span of the Project. Responsibility should be handled as Issue/ acquisition of land should be handled. Issue of Utilities and arrangement of Funds is the sole responsibility of the Authority. One senior officer from Authority and various representative will work at various location under same Authority. At every 50 km two representative of Authority should be available at Engineer's Representative office.

ARCHITECT & ENGINEER- Selection should be a single Joint Venture minimum of two Firm either Government firm or Private or Corporation But not more than 5 Firm. Preferences should be to the Single Institution based on their vision. Example ENGINEER LIVE FOUNDATION whose vision is similar voluntarily work should be given.

Selected FIRM should continue for Life Long for the above work. Responsibility should be handled as Research and Development activity, Design of Work, Costing of the Work, Bidding and finalization of Contractor, Supervision of the Work, Billing and maintenance of the Work. ARCHITECT & ENGINEER has responsibility to run the institution to teach or train the Student regarding Technology involved in Work.sssss

No work should be designed for less than 100 years of Service Life.

Single ENGINEER and various representatives will work at various locations under the same ENGINEER. For every 300 km one Engineer's Representative office with R&D lab, design team and supervision team should be available to handle the Whole Work. Construction Period should be considered as 20 years in 5 stages as per fund arrangement; hence 10 division should be developed. Details are as under-.

R&D Lab - Research and Development is an essential tool to develop the WORK should run by Engineers Representative for Material design, Mix Design, Maintenance material Design and

Product designing at site. R&D design team have responsibility to design the innovative material, mixes and product.

Design Lab - This lab is essential to develop the innovative design suitable for the specific reach or area. Innovative design helps Work to develop. Design Lab have responsibility to develop the design is such that design should run for a long period. Continuous innovative design research is the GOAL.

Costing, Bidding, Billing & Contract Lab - Costing is an innovative and depends on daily market condition, availability of material, technology development, product development and material development. Bidding defining at planning stage and on based on planning stage overhead cost should be calculated for each bidding. Contract is modified as per Planning, material design, product design, innovative design of Work and Bidding. Costing Lab have responsibility to collect the local govt. Taxes, political issue, local material rate, local labor and worker rate, deriving of innovative material rate, innovative product rate, developing overhead calculation, rate analysis for all different items being used in concerned Work or in Bidding at regular interval of each month for the billing and costing purposes. Billing at regular interval should carried for the concerned contractor on the basis of revised rate at interval of every month and contractor bidding concern such as on rebate or on additional rate. Bidding laboratory has responsibility to prepare the bid for the Work and getting concerned rate either in rebate or in addition with desired experience normally 10% work completed within 2 to 3 year and including availability of Manpower, material, machinery and plant availability. Bidding Laboratory has responsibility to collect theEarnest Money Deposit (2% of Value) this is refundable in case work not awarded.

Security deposit or performance Guarantee of 5%-10% value of contract value to submitted at time of start of work in form of Bank Guarantee and same time Earnest Money deposit should return to contractor and Retention money of 5%-0% to be deducted in each bill by Billing Engineer. In case after becoming lowest in bidding if contractor leaving the Contract, EMD should be forfeited by Engineer. In calculation of Change of Scope actual and agreed cost should be paid to contractor as a different Contract in case of short-term work otherwise paid on actual rate and executed quantity. Variation is the responsibility of Engineer and this is essential to do in case of innovative or new technology. Escalation issue will not arise as rate on monthly basisis finalized for the payment. (Utilization of Index is not appropriate to calculate the cost, Index is for calculation of approximate costfor budget purposes)

DISPUTE RESOLUTION LAB- Dispute Resolution Lab should developed by Engineer at Architect and Engineer's office and Engineer's Representative office, where innovative design, product development and material testing and discussion chamber should prepare. For every project 6-member team 2 from each Authority, Engineer & Contractor under leadership of Engineer should auditthe whole laboratory at every 90 days and submitting the final reporton 7th day at site which should be followed by all. In any special dispute this team will visit the project and resolve the issue. In caseContractor is not agreed or Engineer do not understand then threeArbitrator (One from each team) should be appointed by Engineerto resolve the issue in specified time. If Still case not resolved thenamicable settlement, is the only solution. If still contractor is not agreed or Engineer have trouble then Engineer has to take over the contractor and complete the project on time.

HEALTH, SAFETY AND ENVIORNMENT LAB- This is most important for the project as health is the right of every individual, hence health center is essential to establish for every project under Architect & Engineer. Safety is another important which is two types is either personal safety or construction safety is required toestablish in every contract by Contractor and Environmental Lab is the responsibility of the Engineer to establish to check the Environmental check at interval of 3 months and various remedialaction taken by Contractor as required by the Engineer.

Health & Life Insurance and health card of every employee and team member is essential for every project and each individual is eligible to visit the health lab for treatment.

Insurance of Plant and Machinery is essential for the safety.

Live Environmental Board is essential to display theEnvironmental condition of the Project.

LAND ACQUISITION, UTILITY DEVELOPMENT & SHIFTING & CO-ORDINATION LAB- Under Authority Representative this Lab should run the work without hindrances at Engineer's and Engineer's Representative office.

INSTITUTION LAB-Development of Institution is the responsibility of the Engineer to bring the institution in below the 100th rank in world.

EMMERGENCY LAB-Emergency Lab is responsibility of Engineer and Contractor team. All Emergency tools should available in Lab with 24x7x365 days basis responsible team should available in this Lab to tackle the issue after complaint on urgent basis (Immediate basis)

MAINTENANCE LAB - Maintenance Lab is an essential requirement to maintain the Work before and after completion of work. Maintenance work is a continuous activity till dismantle of the Work hence it needs permanent requirement of Maintenance team to work at site, material, machinery, plant, tools etc. handle the issue on Emergency. Following procedure should followed for the Maintenance Work is as under-

1. Daily Inspection and Repair of any fault in Work.
2. Monthly Joint Inspection of all exposed elements of Work and maintaining fault wise.
3. Denting, Painting and repair activity of structure before Rainy season and after rainy season after detailed Inspection.
4. Payment should be based on Cost plus Method. In case Work used by the Public then Public Private Partnership basis.

FUND DEVELOPMENT LAB- Under Authority this Lab runs. Daily wise fund management watched and discussed on the basis of prepared & recommendation of Engineer and Engineer Representative.

Case Study-

Suppose a Ganga Development Corporation developed where 2510km has to develop then whole stretch should divided in 250km, means 10 stretches should prepared. Each one should handle by Engineers representative.250 km should divide in 12 no. stretch of25km each.

INITIAL FUNDING-

1. Office of **ARCHITECT & ENGINEER** is developed after appointment by the Employer of Architect & Engineer where Authority will also accommodate. Dispute Resolution Team is sitting in office under the Architect & Engineer. Library should develop at office. Monthly meeting with Engineer's Representative,

Authority Representative should organize. Online Weekly meeting organized with Engineer's Representative and Authority Representative. Monthly date wise meeting with Contractor Representative, Engineer Representative and Authority Representative should discussed. Central laboratory shoulddevelop at office to minimum design & testing of 1%. Head of Institution should develop at office to run or co-ordination of 10 number Technical Institution at Engineer's Representative office. Expenditure incurred by office should born by the Employer.

2. Offices of ten number Engineer's Representative with Authority Representative should developed with Laboratory for research and development on 24x7 basis, laboratory for innovative material development, laboratory of innovative design, laboratory of Costing, laboratory of Bidding & Contract developed under Engineer's Representative. Laboratory of Land acquisition, Utility development and shifting and co-ordination should develop under the Authority Representative. Either ten institutions attached with Corporation or ten new technical institution should develop under Engineer's Representative. All Architect, Engineer & Managerial team has to take practical class in Institution minimum once. All laboratory should open for 24x7 basis to the student.
3. Initial Funding should be organized by the Authority through taxes or various loans organized by Authority Representative.

PRIMARY FUNDING FOR IST PHASE

4. Construction should be divided in 4 phases a. Ist phase – 5years (5%) b.2nd Phase-10 years (20%) c. 3$^{rd.}$ phase-15 years (50%) and d.4th phase-20 years (100%).5th phase-Maintenance phase to end of life.

5. Primary funding needs for Ist 5 year to complete at least 10 number projects of 25km work. It involves preparation of feasibility report, preliminary costing, detailed research and development, material development, design and DPR preparation costing. Tender documents preparation, selection of Contractor and completing the work.

6. Mostly fund should be organized by the Authority to collect the fund from various taxes and loans.

SECONDARY FUNDING FOR 2ND AND 3RD PHASE

7. After completion of Ist phase various taxes should come directly from petrol pump, shopping area or parking zone, utilization of product, from institution and other means to directly to the Corporation account. Further development needs more money to develop the infrastructure. Authority will further organize from various taxes and loans.

FINAL FUNDING FOR 4TH PHASE AND MAINTENANCE PHASE-

8. After completion of 2nd and $3^{rd.}$ phase Authority will receive the sufficient funds for the 4th stage construction and maintenance of 1st,2nd and 3rd phase work and repayment of various loans.

SURPLUS FUNDING-

9. After completion of 4th phase of work sufficient funds should receive from the area from petrol pump, market and other taxesand fees from institution and services. Expenditure includes the repayment of loans and taxes and in case of sufficient funding loan should be given to the other infrastructure development or work.

CONTRACTOR - Contractor Should differ for different work as described in BIDDING. Contractor will differ from Project to Project, but Contractor should continue for execute projects till End of Service Life for the same project from Execution to Maintenance.

Generally, for every 25 km for assumed corporation one contractor should be deployed to execute the work in the right way. Work Program, Time of Extension, Work Methodology or any other important issue, contractor has to present to the Engineer atEngineers office and approval taken from Engineer at end of Presentation.

EMERGENCY WORK - Separate Contractors should be deployed as per work and time suitability.

DISPUTE RESOLUTION COMMITTEE - Dispute resolution committee is the team of Expert of Authority, Engineer and Contractor whose age should be 65 yrs. to 90 yrs. At every 90 days 7-day team of dispute resolution of 6 member (Two from each) should visit all activity under Leadership of ENGINEER and will comment jointly on spot on 7th day before leaving the site, Comment has to follow by the Team. Without report team cannot leave the Work Site.

DISAGREEMENT - Disagreement between two parties should be resolved by the Committee and as per joint report action should be taken.

Takeover of Contractor - As Democratic point of view any party who joined the hand of GOVT should not be out in any case. In worst condition of Contractor should be handed over by Engineer for Employer and Firm should be run by different team deployedby the Employer. After completing join Firm may sold out with employee to another Firm.

Termination word is a danger for Democratic Contract

Example - Suppose the Contractor is failing to execute the work due to justified financial issues or else, with recommendation of the

COMMITTEE, Government has to take suitable action, such that Project Work should not suffer. In only extraordinary and exceptional condition, no way Contractor should be terminated.

Employment of Foreign Nationals - The Contractor and Engineer acknowledges, agrees and undertakes that employment of foreign personnel by the Contractor and/or its Subcontractors andtheir sub-contractors shall be subject to grant of requisite regulatory permits and approvals including employment/ residential visas and work permits, if any required as per governments norms, and the obligation to apply for and obtain thesame shall always rest with the Engineer & Contractor.Notwithstanding anything to the contrary contained in this Agreement, refusal of or inability to obtain any such permits and approvals by the Contractor or any of its Subcontractors or their subcontractors, shall not constitute a Force Majeure Event, and shall not in any manner excuse the Contractor from the performance and discharge of its obligations and liabilities underthis Agreement, and the Contractor's liabilities hereunder shall remain unaffected by such failure, refusal or inability.

Employment of Foreign nationals should be for the time of 1-10 yrs. only.

The forms are recommended for general use where tenders are invited on an international basis. Modifications may be required in some jurisdictions, particularly if the Conditions are to be used on domestic contracts. ELF considers the official and authentic textsto be the versions in the English language.

In the preparation of these Conditions of Contract, it wasrecognized that, while there are many sub-clauses which will be generally applicable, there are some sub-clauses which must necessarily vary to take account of the circumstances. relevant to the particular contract. The sub-clauses which were considered to be applicable to many (but not all) contracts have been included in the General Conditions, in order to facilitate their incorporation into each contract.

The General Conditions and the Particular Conditions will together comprise the Conditions of Contract governing the rights and obligations of the parties. It will be necessary to prepare the Particular Conditions for each individual contract, and to take account of those sub-clauses in the General Conditions which mention the Particular Conditions.

For this publication, the General Conditions were prepared on the following basis:

i. Interim and final payments will be determined by measurement, applying the rates and prices in a Bill of Quantities.

ii. If the wording in the General Conditions necessitates further data, then (unless it is so descriptive that it would have to be detailed in the Specification) the sub - clause makes referenceto this data being contained in the Appendix to Tender, the data either being prescribed by the Engineer or being insertedby the Tenderer.

iii. Where a sub-clause in the General Conditions deals with a matter on which different contract terms are likely to be applicable for different contracts, the principles applied in writing the sub-clause were:

a) users would find it more convenient if any provisions which they did not wish to apply could simply be deleted or not invoked, than if additional text had to be written (in the Particular Conditions) because the General Conditions did notcover their requirements; or

b) In other cases, where the application of (a) was thought to be inappropriate, the sub-clause contains the provisions which were considered applicable to most contracts.

For example, Sub-Clause 14.2 [Advance Payment] is included for convenience, not because of any ELF Democratic Contract policyin respect of advance payments. This Sub-Clause becomes inapplicable (even if it is not deleted) if it is disregarded by not specifying the amount of the advance. It should therefore be notedthat some of the provisions contained in the General Conditions may not be appropriate for an apparently typical contract.

Further information on these aspects, example wording for other arrangements, and other explanatory material and example wording to assist in the preparation of the Particular Conditions and the other tender documents, are included within this publication as Guidance for the Preparation of the ParticularConditions. Before incorporating any example wording, it must bechecked to ensure that it is wholly suitable for the particular circumstances; if not, it must be amended.

Where example wording is amended, and in all cases where other amendments or additions are made, care must be taken to ensure that no ambiguity is created, either with the General Conditions orbetween the clauses in the Particular Conditions. It is essential thatall these drafting tasks, and the entire preparation of the tender documents, are entrusted to personnel with the relevant expertise, including the contractual, technical and procurement aspects.

This publication concludes with example forms for the Letter of Tender, the Appendix to Tender (providing a checklist of the sub-clauses which refer to it), the Contract Agreement, and alternativesfor the Dispute Adjudication Agreement. This Dispute Adjudication

Agreement provides text for the agreement between the **AUTHORITY, ARCHITECT & ENGINEER AND CONTRACTOR** and the person appointed to act either as sole adjudicator or as a member of a SIX-person dispute adjudication board; and incorporates (by reference) the terms in the Appendix to the General Conditions.

ELF intends to publish a guide to the use of its Conditions of Contract for Infrastructure Development Construction, Maintenance and Emergency Work. Another relevant ELF publication is "Tendering Procedure", which presents a systematicapproach to the selection of tenderers and the obtaining andevaluation of tenders.

In order to clarify the sequence of Contract activities, reference may be made to the charts on the next two pages and to the Sub- Clauses listed below (some Sub-Clause numbers are also stated inthe charts). The charts are illustrative only and must not be taken into consideration in the interpretation of the Conditions of Contract.

1.1.3.1 &	13.7	Base Date
1.1.3.2 &	8.1	Commencement Date
1.1.6.6 &	4.2	Performance Security
1.1.4.7 &	14.3	Interim Payment Certificate
1.1.3.3 &	8.2	Time for Completion (as extended under 8.4)

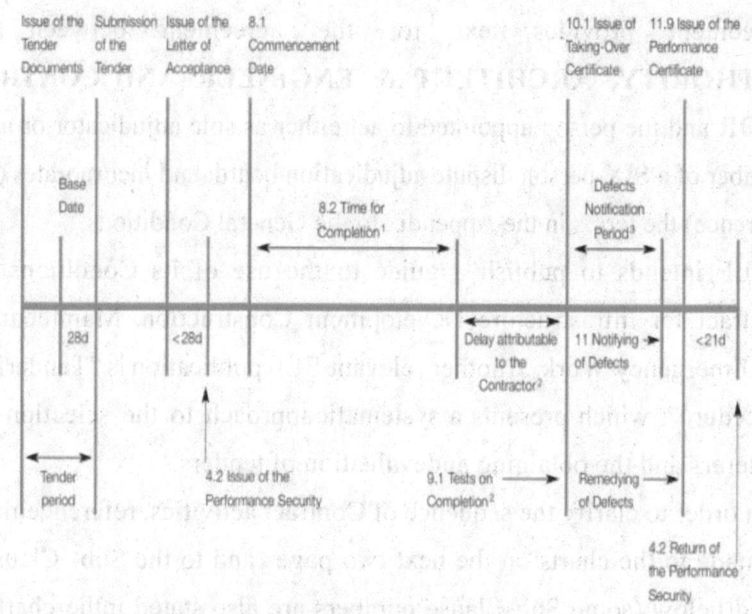

1.1.3.5 &	10.1	Taking-Over Certificate
1.1.3.7&11.3)	11.1	Defects Notification Period (as extended under
1.1.3.8 &	11.9	Performance Certificate
1.1.4.4 &	14.13	Final Payment Certificate

Typical sequence of Principal Events during Contracts for Construction

a) The Time for Completion is to be stated (in the Appendix to Tender) as a number of days, to which is added any extensions of time under Sub-Clause 8.4.

b) In order to indicate the sequence of events, the above diagram is based upon the example of the Contractor failing to comply with Sub-Clause 8.2.

c) The Defects Notification Period is to be stated (in the Appendix to

Tender) as a number of days, to which is added any extensions under Sub-Clause 11.3

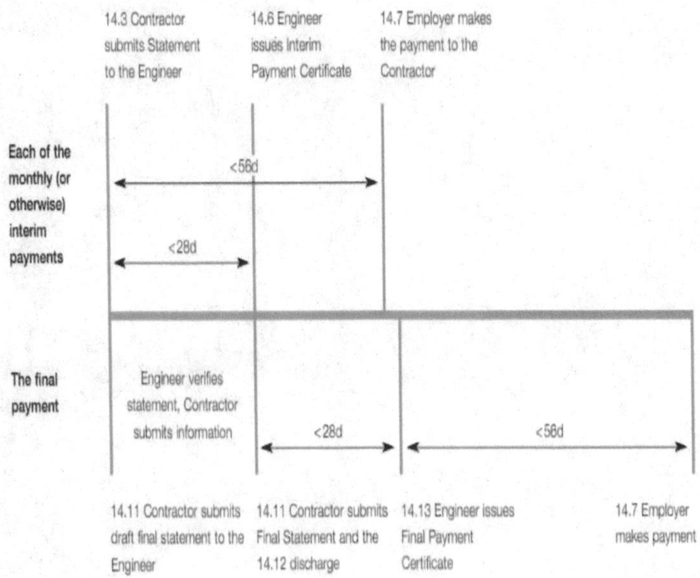

Typical sequence of Payment Events envisaged in Clause 14

Typical sequence of Dispute Events envisaged in Clause 20GENERAL CONDITIONS

GUIDANCE FOR THE PREPARATION OF PARTICULAR CONDITIONS

FORMS OF LETTER OF TENDER,

CONTRACT AGREEMENT AND DISPUTE ADJUDICATION AGREEMENT

Conditions of Contract for CONSTRUCTION

FOR INFRASTR UCTURE DEVELOPMENT CONTRACT

Contents

Appendix

GENERAL CONDITIONS OF DISPUTE ADJUDICATION AGREEMENT

SCHEDULE

INDEX OF SUB-CLAUSES

1.1.6.3 Employer's Equipment

1.1.2.6 Employer's Personnel

1.1.2.4 Engineer

1.1.2.10 FIDIC

1.1.4.4 Final Payment Certificate

1.1.4.5 Final Statement

1.1.6.4 Force Majeure

1.1.4.6 Foreign Currency

1.1.5.2 Goods

1.1.4.7 Interim Payment Certificate

1.1.6.5 Laws

1.1.13 Letter of Acceptance

1.1.1.4 Letter of Tender

1.1.4.8 Local Currency

1.1.5.3 Materials

1.1.2.1 Party

1.1.4.9 Payment Certificate

1.1.3.8 Performance Certificate

1.1.6.6 Performance Security

1.1.5.4 Permanent Works

1.1.5.5 Plant

1.1.4.1 0Provisional Sum

1.1.4.10Retention Money

1.1.1.7 Schedules

1.1.5.6 Section

1.1.6.7 Site

1.1.1.5 Specification

1.1.4.12 Statement

1.1.2.8 Subcontractor

1.1.3.5 Taking-Over Certificate

1.1.5.7 Temporary Works

1.1.1.8 Tender

1.1.3.6 Tests after Completion

1.1.3.4 Tests on Completion

1.1.3.3 Time for Completion

1.1.6.8 Unforeseeable

1.1.6.9 Variation

1.1.5.8 Works

1.1.3.9 year

Part I Preliminary

General Provision

1.1 Definition

In the Conditions of Contract ("these Conditions"), which include Particular Conditions and these General Conditions, the following words and expressions shall have the meanings stated. Words indicating persons or parties include corporations and other legal entities, except where the context requires otherwise.

1.1.1 The Contract

1.1.1.1 **"Contract"** means the Contract Agreement, the Letter of Acceptance, the Letter of Tender, these Conditions, the Specification, the Drawings, the Schedules, and the further documents (if any) which are listed in the Contract Agreement or in the Letter of Acceptance.

1.1.1.2 "**Contract Agreement"** means the contract agreement (if any) referred to in Sub-Clause 1.6 [Contract Agreement].

1.1.1.3 "**Letter of Acceptance**" means the letter of formal acceptance, signed by the Engineer, of the Letter of Tender, including any annexed memoranda comprising agreementsbetween and signed by both Parties. If there is no such letter of acceptance, the expression "Letter of Acceptance" means the Contract Agreement and the date of issuing or receiving the Letter of Acceptance means the date of signing the Contract Agreement.

1.1.1.4 **"Letter of Tender"** means the document entitled letter of tender, which was completed by the Contractor and includes the signed offer to the Employer for the Works.

1.1.1.5 **"Specification"** means the document entitled specification, as included in the Contract, and any additions and modifications to the specification in according to the Contract. Such a document specifies the Works.

1.1.1.6 **"Drawings"** means the drawings of the Works, as included in the Contract, and any additional and modified drawings issuedby (or on behalf of) the Employer in accordance with the Contract.

1.1.1.7 **"Schedules"** means the document(s) entitled schedules, completed by the Contractor and submitted with the Letter of Tender, as included in the Contract. Such document may include the Bill of Quantities, data, lists, and schedules of rates and/or prices.

1.1.1.8 **"Tender"** means the Letter of Tender and all other documents which the Contractor submitted with the Letter of Tender, as included in theContract.

1.1.1.9 **"Appendix to Tender"** means the completed pages entitled appendix to tender which are appended to and form part of the Letter of Tender.

1.1.1.10 **"Bill of Quantities"** and **"Daywork Schedule"** mean the documents so named (if any) which are included in the Schedules.

1.1.2 Parties and Person

1.1.2.1 **"Party"** means the Employer, Authority, ENGINEER or the Contractor, as the context requires.

1.1.2.2.1 **"Democratic Government/Democratic Employer's"** means who appoints the Authority and Engineer for Corporation/Departments or Infrastructure development or Plant setup or Maintenance Work.

Democratic Government/Employers are the representative of the

People, that means for India the Government is **PRESIDENT** andfor the State Government it is **GOVERNOR.**

1.1.2.1.2 **"Authority"** means firm and representative of specific Corporation or Department representatives appointed by the Government. Their role is Clearing and arranging of Land and Utility and arranging of Funds. Act as a Coordinator with other Corporation and Departments.

1.1.2.3 "**Contractor**" means the person(s) named as contractor in the Letter of Tender accepted by the Engineer and the legal successors in title to this person(s).

1.1.2.4 **"Architect & Engineer"** means the **ENGINEERING FIRM** appointed by the GOVERNMENT to act as the Architect & Engineer for the purposes of the Contract and named in the Appendix to Tender, and competent person appointed from time to time by theENGINEERING FIRM and notified to the Employer's, AUTHORITY & Contractor under Sub-Clause 3.4 [Replacement of the Engineer].

1.1.2.5 **"Contractor's Representative"** means the person named by the Contractor in the Contract or appointed from time to time by the Contractor under Sub-Clause 4.3 [Contractor's Representative], who acts on behalf of the Contractor.

1.1.2.6 **"Employer's Personnel"** means the Authority & Engineer, the assistants referred to in Sub-Clause 3.2 [Delegation by the Engineer] and all other staff, labor and other employees of the Engineer and of the Employer; and any other personnel notified to the Contractor, by the Engineer, as Employer's Personnel.

1.1.2.7 **"Contractor's Personnel"** means the Contractor's

Representative and all personnel whom the Contractor utilizes onSite, who may include the staff, labor and other employees of the Contractor and of each Subcontractor; and any other personnel assisting the Contractor in the execution of the Works.

1.1.2.8 **"Subcontractor"** means any person named in the Contract as a subcontractor, or any person appointed as a subcontractor, fora part of the Works; and the legal successors in title to each of these persons.

1.1.2.9 **"DAB"** means the person, or SIX persons so named in the Contract, or other person(s) appointed under Sub-Clause 20.2 [Appointment of the **Dispute Adjudication Board**] or Sub-Clause 20.3 [Failure to Agree DisputeAdjudication Board]

1.1.2.10 **"ELF"** means the Engineer live Foundation, the creator of Democratic Contract

1.1.3 Dates, Tests, Periods and Completion

1.1.3.1 "Base Date" means the date 28 days prior to the latest date for submission of the Tender.

1.1.3.2 **"Commencement Date"** means the date notified under Sub-Clause 8.1 [Commencement of Works].

1.1.3.3 **"Time for Completion"** means the time for completing the Works or a Section (as the case may be) under Sub-Clause 8.2 [Time for Completion], as stated in the Appendix to Tender (with any extension under Sub Clause 8.4 [Extension of Time for Completion]), calculated from the Commencement Date.

1.1.3.4 **"Tests on Completion"** means the tests which are specified in the Contract or agreed by both Parties or instructed as aVariation, and which are carried out

1.1.3.5 **"Taking-Over Certificate"** means a certificate issued under Clause 10 [Employer's Taking Over].

1.1.3.6 **"Tests after Completion"** means the tests (if any) whichare specified in the Contract and which are carried out in accordance with the provisions of the Particular Conditions after the Works or a Section (as the case may be) are taken over by the Employer.

1.1.3.7 **"Defects Notification Period"** means the period for notifying defects in the Works or a Section (as the case may be) under Sub-Clause 11.1

[Completion of Outstanding Work and Remedying Defects], as stated in the Appendix to Tender (with any extension under Sub- Clause 11.3 [Extension of Defects Notification Period]), calculated from the date on which the Works or Section is completed as certified under Sub-Clause 10.1 [Taking Over of the Works and Sections].

1.1.3.8 **"Performance Certificate"** means the certificate issued under Sub Clause 11.9 [Performance Certificate].

1.1.3.9 **"day"** means a calendar day and **"year"** means 365 days.

1.1.4 Money and Payments

1.1.4.1 **"Accepted Contract Amount"** means the amount acceptedin the Letter of Acceptance for the execution and completion of the Works and theremedying of any defects.

1.1.4.2 **"Contract Price"** means the price defined in Sub-Clause

14.1 [The Contract Price] and includes adjustments in accordance with the Contract.

1.1.4.3 **"Cost"** means all expenditure reasonably incurred (or to be incurred) by the Contractor, whether on or off the Site, including overhead and similar charges, but does not include profit.

1.1.4.4 **"Final Payment Certificate"** means the payment certificate issued under Sub-Clause 14.13 [Issue of Final Payment Certificate].

1.1.4.5 **"Final Statement"** means the statement defined in Sub-Clause 14.11 [Application for Final Payment Certificate].

1.1.4.6 **"Foreign Currency"** means a currency in which part (or all)of the Contract Price is payable, but not the Local Currency.

1.1.4.7 **"Interim Payment Certificate"** means a payment certificate issued under Clause 14 [Contract Price and Payment], other than the Final Payment Certificate.

1.1.4.8 **"Local Currency"** means the currency of the Country.

1.1.4.9 **"Payment Certificate"** means a payment certificate issued under Clause 14 [Contract Price and Payment].

1.1.4.10 **"Provisional Sum"** means a sum (if any) which is specifiedin the Contract as a provisional sum, for the execution of any partof the Works or for the supply of Plant, Materials or services under Sub-Clause 13.5 [Provisional Sums].

1.1.4.11 **"Retention Money"** means the accumulated retention moneys which the Employer retains under Sub-Clause 14.3 [Application for Interim Payment Certificates] and pays under Sub-Clause 14.9 [Payment of Retention Money].

1.1.4.12 **"Statement"** means a statement submitted by the Contractor as part of an application, under Clause 14 [ContractPrice and Payment], for a payment Certificate.

1.1.5 Works and Goods

1.1.5.1 **"Contractor's Equipment"** means all apparatus, machinery, vehicles and other things required for the execution and completion of the Works and the remedying any defects. However, Contractor's Equipment excludes Temporary Works, Employer'sEquipment (if any), Plant, Materials and any other things intended to form or forming part of the Permanent Works.

1.1.5.2 **"Goods"** means Contractor's Equipment, Materials, Plantand Temporary Works, or any of them as appropriate.

1.1.5.3 **"Materials"** means things of all kinds (other than Plant) intended to form or forming part of the Permanent Works, including the supply-only materials (if any) to be supplied by the Contractor under the Contract.

1.1.5.4 **"Permanent Works"** means the permanent works to be executed by the Contractor under the Contract.

1.1.5.5 **"Plant"** means the apparatus, machinery and vehicles intended to form or forming part of the Permanent Works.

1.1.5.6 **"Section"** means a part of the Works specified in the Appendix to Tender as a Section (if any).

1.1.5.7 **"Temporary Works"** means all temporary works of everykind (other than Contractor's Equipment) required on Site for the execution and completion of the Permanent Works and the remedying of any defects.

1.1.5.8 **"Works"** mean the Permanent Works and the Temporary Works, or either of them as appropriate.

1.1.6 Other Definitions

1.1.6.1 **"Contractor's Documents"** means the calculations, computer programs and other software, drawings, manuals, models and other documents of a technical nature (if any) supplied by the Contractor under the Contract.

1.1.6.2 **"Country"** means the country in which the Site (or most ofit) is located, where the Permanent Works are to be executed.

1.1.6.3 **"Employer's Equipment"** means the apparatus, machinery and vehicles (if any) made available by the Employer for the use of the Contractor in the execution of the Works, as stated in the Specification; but does not include Plant which has not been taken over by the Employer.

1.1.6.4 **"Force Majeure"** is defined in Clause 19 [Force Majeure].

1.1.6.5 **"Laws"** means all national (or state) legislation, statutes, ordinances and other laws, and regulations and by-laws of any legally constituted public Authority.

1.1.6.6 **"Performance Security"** means the security (or securities,if any) under Sub-Clause 4.2 [Performance Security].

1.1.6.7 **"Site"** means the places where the Permanent Works are tobe executed and to which Plant and Materials are to be delivered, and any other places as may be specified in the Contract as forming part of the Site.

1.1.6.8 **"Unforeseeable"** means not reasonably foreseeable by an experienced contractor by the date for submission of the Tender.

1.1.6.9 **"Variation"** means any change to the Works, which is instructed or approved as a variation under Clause 13 [Variationsand Adjustments].

1.2 Interpretation

In the Contract, except where the context requires otherwise:

a. Words indicating one gender include all genders.

b. Words indicating the singular also include the plural and words indicating the plural also include the singular.

c. Provisions including the word "agree", "agreed" or "agreement" require the agreement to be recorded in writing, and

d. "written" or "in writing" means hand-written, type-written, printed or electronically made, and resulting in a permanent record.

The marginal words and other headings shall not be taken into consideration in the interpretation of these Conditions.

1.3 Communications

Wherever these Conditions provide for the giving or issuing ofapprovals, certificates, consents, determinations, notices and requests, these communications shall be:

a. in writing and delivered by hand (against receipt), sent by mail or courier, or transmitted using any of the agreed systems of electronic transmission as stated in the Appendix to Tender; and

b. delivered, sent or transmitted to the address for the recipient's communications as stated in the Appendix to Tender. However:

i. if the recipient gives notice of another address, communications shall thereafter be delivered; accordingly, and

ii. if the recipient has not stated otherwise when requesting an approval or consent, it may be sent to the ad dress from whichthe request was Issued.

Approvals, certificates, consents and determinations shall not be unreasonably withheld or delayed. When a certificate is issued to aParty, the certifier shall send a copy to the other Party. When a notice is issued to a Party, by the other Party or the Engineer, a copy shall be sent to the Engineer or the other Party, as the case may be.

1.4 Law and Language

The Contract shall be governed by the law of the country (or another jurisdiction) stated in the Appendix to Tender.

If there are versions of any part of the Contract which are writtenin more than one language, the version which is in the ruling language stated in the Appendix to Tender shall prevail.

The language for communications shall be that stated in the Appendix to Tender. If no language is stated there, the languagefor communications shall be the language in which the Contract (or most of it) is written.

1.5 Priority of Documents

The documents forming the Contract are to be taken as mutually explanatory of one another. For the purposes of interpretation, the priority of the documents shall be in accordance with the following sequence:

a. the Contract Agreement (if any),
b. the Letter of Acceptance,
c. the Letter of Tender,
d. the Particular Conditions,
e. these General Conditions,
f. the Specification,
g. the Drawings, an
h. the Schedules and any other documents forming part of the Contract.

If an ambiguity or discrepancy is found in the documents, the Engineer shall issue any necessary clarification or instruction.

1.6 Contract Agreement

The Parties shall enter into a Contract Agreement within 28 days after the Contractor receives the Letter of Acceptance, unless theyagree otherwise.

The Contract Agreement shall be based upon the form annexed tothe Particular Conditions.

The costs of stamp duties and similar charges (if any) imposed by law in connection with entry into the Contract Agreement shall be borne by the ENGINEER on behalf of the Employer.

1.7 Assignment

Neither Party shall assign the whole or any part of the Contract orany benefit or interest in or under the Contract to anyone else. However, either Party:

a) may assign the whole or any part with the prior agreement ofthe other Party, at the sole discretion of such other Party, and
b) may, as security in favor of a bank or financial institution, assign its right to any money due, or to become due, under theContract.

1.8 Care and Supply of Documents

The Specification and Drawings shall be in the custody and care ofthe Engineer. Unless otherwise stated in the Contract, two copiesof the Contract and of each subsequent Drawing shall be suppliedto the Contractor, who may make or request further copies at the cost of the Contractor.

Each of the Contractor's Documents shall be in the custody and care of the Contractor, unless and until taken over by the Employer. Unless otherwise stated in the Contract, the Contractorshall supply to the Engineer six copies of each of the Contractor'sDocuments.

1.9 Delayed Drawings or Instructions

The Contractor shall keep on the Site, a copy of the Contract, publications named in the Specification, the Contractor's Documents (if any), the Drawings and Variations and other communications given under the Contract. The Engineer's Personnel shall have the right of access to all these documents atall reasonable times.

If a Party becomes aware of an error or defect of a technical naturein a document which was prepared for use in executing the Works,the Party shall promptly give notice to the other Party of such error or defect.

The Contractor shall give notice to the Engineer whenever the Works are likely to be delayed or disrupted if any necessary drawing or instruction is not issued to the Contractor within a particular time, which shall be reasonable. The notice shall include details of the necessary drawing or instruction, details of why andby when it should be issued, and details of the nature and amountof the delay or disruption likely to suffer if it is late.

If the Contractor suffers delay and/or incurs Cost as a result of a failure of the Engineer to issue the notified drawing or instructionwithin a time which is reasonable and is specified in the notice with supporting details, the Contractor shall give a further notice to theEngineer and shall be entitled subject to Sub-Clause 20.1 [Contractor's Claims] to:

a) an extension of time for any such delay, if completion is or will be delayed, under Sub-Clause 8.4 [Extension of Time for Completion], and
b) payment of any such Cost-plus reasonable profit, which shallbe included in the Contract Price. After receiving this furthernotice, the Engineer shall proceed in accordance with Sub Clause 3.5 [Determinations] to agree or determine these matters. However, if

and to the extent that the Engineer's failure was caused by any error or delay by the Contractor, including an error in, or delay in the submission of, any of the Contractor's Documents, the Contractor shall not be entitled to such extension of time, Cost or profit.

1.10 Employer's Use of Contractor's Documents

As between the Parties, the Contractor shall retain the copyright and other intellectual property rights in the Contractor's Documents and other design documents made by (or on behalf of)the Contractor.

The Contractor shall be deemed (by signing the Contract) to giveto the Engineer a non-terminable transferable non-exclusive royalty-free license to copy, use and communicate the Contractor's Documents, including making and using modifications of them. This license shall:

a) apply throughout the actual or intended working life (whichever is longer) of the relevant parts of the Works,

b) entitle any person in proper possession of the relevant part ofthe Works to copy, use and communicate the Contractor's Documents for the purposes of completing, operating, maintaining, altering, adjusting, repairing and demolishing the Works, and

c) in the case of Contractor's Documents which are in the form of computer programs and other software, permit their use on any computer on the Site and other places as envisaged by theContract, including replacements of any computers supplied by the Contractor.

The Contractor's Documents and other design documents made by(or on behalf of) the Contractor shall not, without the Contractor'sconsent, be used, copied or communicated to a third party by (or on behalf of) the Employer for purposes other than those permitted under this Sub-

Clause.

1.11 Contractor's Use of Engineer's Documents

As between the Parties, the Engineer shall retain the copyright and other intellectual property rights in the Specification, the Drawings and other documents made by (or on behalf of) the Engineer. The Contractor may, at his cost, copy, use, and obtain communication of these documents for the purposes of the Contract. They shall not, without the Engineer's consent, be copied, used or communicated to a third party by the Contractor, except as necessary for the purposes of the Contract.

1.12 Confidential Details

The Contractor shall disclose all such confidential and other information as the Engineer may reasonably require in order to verify the Contractor's compliance with the Contract.

1.13 Compliance with Laws

The Contractor shall, in performing the Contract, comply with applicable Laws. Unless otherwise stated in the Particular Conditions:

a) the Engineer shall have obtained (or shall obtain) the planning, zoning or similar permission for the Permanent Works, and any other permissions described in the Specification as havingbeen (or being) obtained by the Engineer; and the Engineer shall indemnify and hold the Contractor harmless against and from the consequences of any failure to do so; and

b) the Contractor shall give all notices, pay all taxes, duties and fees, and obtain all permits, licenses and approvals, as required by the Laws in relation to the execution and completion of the Works and the remedying of any defects; and the Contractor shall indemnify and hold the Engineer harmless against and from the consequences of any failure to do so.

1.14 Joint and Several Liability

If the Contractor constitutes (under applicable Laws) a joint venture, consortium or other unincorporated grouping of two or more persons:

a) these persons shall be deemed to be jointly and severally liable to the Employer for the performance of the Contract.

b) these persons shall notify the Employer of their leader who shall have authority to bind the Contractor and each of these persons; and

c) the Contractor shall not alter its composition or legal status without the prior consent of the Employer.

PART II SCOPE OF DEVELOPMENT

Under this Agreement, the scope of the DEVELOPMENT (the "Scope of the Project") shall mean and include:

a) Construction/Maintenance of the DEVELOPMENT on the Site set forth in Schedule- A and as specified in Schedule-B together with provision of Project Facilities as specified in Schedule-C, and in conformity with the Specifications and Standards set forth in Schedule-D;

b) Maintenance of the DEVELOPMENT in accordance with the provisions of this Agreement and in conformity with the requirements set forth in Schedule-E; and

c) Performance and fulfilment of all other obligations of the Contractor in accordance with the provisions of this Agreement and matters incidental thereto or necessary for the performance of any or all of the obligations of the Contractor under this Agreement.

The Employer/Government

2.1 Right of Access to the Site the Employee

The Engineer shall give the Contractor right of access to, and possession of, all parts of the Site within the time (or times) statedin the Appendix to Tender. The right and possession may not be exclusive to the Contractor. If, under the Contract, the Engineer is required to give (to the Contractor) possession of any foundation, structure, plant or means of access, the Engineer shall do so in the time and manner stated in the Specification. However, the Engineer may withhold any such right or possession until the Performance Security has been received.

If no such time is stated in the Appendix to Tender, the Engineer shall give the Contractor right of access to, and possession of, theSite within such times as may be required to enable the Contractorto proceed in accordance with the program me submitted under Sub-Clause 8.3 [Program me].

If the Contractor suffers delay and/or incurs Cost as a result of a failure by the Engineer to give any such right or possession withinsuch time, the Contractor shall give notice to the Engineer and shall be entitled subject to Sub-Clause 20.1 [Contractor's Claims] to:

a) an extension of time for any such delay, if completion is or will be delayed, under Sub-Clause 8.4 [Extension of Time for Completion], and
b) payment of any such Cost-plus reasonable profit, which shall be included in the Contract Price.

After receiving this notice, the Engineer shall proceed in accordance with Sub Clause 3.5 [Determinations] to agree or determine these matters.

However, if and to the extent that the Engineer's failure was caused by any error or delay by the Contractor, including an errorin, or delay in the submission of, any of the Contractor's Documents, the Contractor shall not be entitled to such extensionof time, Cost or profit.

2.2 Permits, Licenses or Approvals

The Employer/Government shall (where he is in a position to do so) provide reasonable assistance to the Contractor at the request of the Engineer:

a. by obtaining copies of the Laws of the Country which are relevant to the Contract but are not readily available, and

b. for the Contractor's applications for any permits, licenses or approvals required by the Laws of the Country:

i. which the Contractor is required to obtain under Sub-Clause 1.13 [Compliance with Laws],

ii. for the delivery of Goods, including clearance through customs,and

iii. for the export of Contractor's Equipment when it is removedfrom the Site.

2.3 Employer's Personnel

The Employer/Government shall be responsible for ensuring thatthe Employer's Personnel, Authority, Engineer and the Employer's other contractors on the Site:

a) co-operate with the Contractor's efforts under Sub-Clause 4.6[Co-operation], and

b) take actions similar to those which the Contractor is requiredto take under subparagraphs (a), (b) and (c) of Sub-Clause 4.8[Safety Procedures] and under Sub-Clause 4.18 [Protection of the

Environment].

2.4 Employer's Financial Arrangements

The Employer shall submit, within 28 days after receiving any request from the Engineer for Contractor, reasonable evidence that financial arrangements have been made and are being maintained which will enable the Employer to pay the Contract Price (as estimated at that time) in accordance with Clause 14 [Contract Price and Payment]. If the Employer intends to make any materialchange to his financial arrangements, the Employer shall give notice to the Engineer with respect of concerned Contractor with detailed particulars.

2.5 Employer's Claims

If the Employer/Government considers himself to be entitled to any payment under any Clause of these Conditions or otherwise in connection with the Contract, and/or to any extension of the Defects Notification Period, the Engineer shall give notice and particulars to the Contractor. However, notice is not required for payments due under Sub-Clause 4.19 [Electricity, Water and Gas],under Sub-Clause 4.20 [Employer's Equipment and Free-Issue Material], or for other services requested by the Contractor. The notice shall be given as soon as practicable after the Employer became aware of the event or circumstances giving rise to the claim. A notice relating to any extension of the Defects Notification Period shall be given before the expiry of such period.

The particulars shall specify the Clause or other basis of the claim,and shall include substantiation of the amount and/or extension towhich the Employer/Government considers himself to be entitledin connection with the Contract. The Engineer shall then proceedin accordance with

Sub-Clause 3.5 [Determinations] to agree or determine (I) the amount (if any) which the Engineer is entitled tobe paid by the Contractor, and/or (ii) the extension (if any) of theDefects Notification Period in accordance with Sub Clause 11.3 [Extension of Defects Notification Period]. This amount may be included as a deduction in the Contract Price and Payment Certificates. The Engineer shall only be entitled to set off againstor make any deduction from an amount certified in a Payment Certificate, or to otherwise claim against the Contractor, in accordance with this Sub-Clause.

The Architect & Engineer

3.1 Architect & Engineer's Duties and Authority

The Employer/Government shall appoint the Architect & Engineer who shall carry out the duties assigned to him in the Contract.

The Engineer's staff shall include suitably qualified engineers and other professionals who are competent to carry out these duties.

The Engineer shall have authority to amend the Contract after meeting with the Government.

The Engineer may exercise the Employer's attributable to the Engineer as specified in or necessarily to be implied from the Contract.

However, whenever the Engineer exercises a specified authority for which the Employer's approval is required, then (for thepurposes of the Contract) the Employer/Government shall bedeemed to have given approval. Except as otherwise stated in these Conditions:

a) Whenever carrying out duties or exercising authority,specified in or implied by the Contract, the Engineer shall bedeemed to act for the Employer

b) The Engineer has no authority to relieve either Party of any duties, obligations or responsibilities under the Contract; and

c) Any approval, check, certificate, consent, examination, inspection, instruction, notice, proposal, request, test, or similar act by the Engineer (including absence of disapproval) shall not relieve the Contractor from any responsibility he has under the Contract, including responsibility for errors, omissions, discrepancies and non-compliances.

3.2 Delegation by the Architect & Engineer

The Architect & Engineer may from time-to-time assign duties and delegate authority to assistants and may also revoke such assignment or delegation.

These assistants may include a resident engineer, and/or independent inspectors appointed to inspect and/or test items of Plant and/or Materials.

The assignment, delegation or revocation shall be in writing and shall not take effect until copies have been received by both Parties. However, unless otherwise agreed by both Parties, the Engineer shall not delegate the authority to determine any matter in accordance with Sub-Clause 3.5 [Determinations].

Assistants shall be suitably qualified persons, who are competent to carry out these.

3.1 Architect & Engineer's Duties and Authority

3.2 Delegation by the Architect & Engineer the Engineer 3 General Conditions 11 duties and exercise this authority, and who are fluent in the language for communications defined in Sub- Clause 1.4 [Law and Language].

Each assistant, to whom duties have been assigned or authority has been delegated, shall only be authorized to issue instructions to the Contractor to the extent defined by the delegation.

Any approval, check, certificate, consent, examination, inspection, instruction, notice, proposal, request, test, or similar act by an assistant, in accordance with the delegation, shall have the same effect as though the act had been an act of the Engineer. However:

a. any failure to disapprove any work, Plant or Materials shall not constitute approval, and shall therefore not prejudice the right of the Engineer to reject the work, Plant or Materials;

b. if the Contractor questions any determination or instruction ofan assistant, the Contractor may refer the matter to the Engineer, who shall promptly confirm, reverse or vary the determination or instruction.

3.3 Instructions of the Engineer

The Engineer may issue to the Contractor (at any time) instructions and additional or modified Drawings which may be necessary for the execution of the Works and the remedying of any defects, all in accordance with the Contract.

The Contractor shall only take instructions from the Engineer, or from an assistant to whom the appropriate authority has been delegated under this Clause.

If an instruction constitutes a Variation, Clause 13 [Variations and Adjustments] shall apply.

The Contractor shall comply with the instructions given by the Engineer or delegated assistant, on any matter related to theContract. Whenever practicable, their instructions shall be given in writing. If the Engineer or a delegated assistant:

a) gives an oral instruction,

b) receives a written confirmation of the instruction, from (or on behalf of) the Contractor, within two working days after giving the instruction, and

c) does not reply by issuing a written rejection and/or instruction within two working days after receiving theconfirmation, then the confirmation shall constitute the written instruction of the Engineer or delegated assistant (as the case may be).

3.4 Replacement of the Engineer

If the Employer/Government intends to transfer the any Engineerin other project concerned with same Authority in case of misconduct or dishonesty and after recommendation of committee of Dispute Redressal team at every 90 days of 7 days audit, the Engineer shall, not less than 42 days before the intended date of replacement, give notice to the Contractor of the name, address and relevant experience of the intended replacement Engineer.

The Employer shall not replace the Engineer with a person against whom the Contractor raises reasonable objection by notice to the Employer, with supporting particulars.

3.5 Determinations

Whenever these Conditions provide that the Engineer shall proceed in accordance with this Sub-Clause 3.5 to agree or determine any matter, the Engineer shall consult with each Party in an endeavor to reach agreement.

If agreement is not achieved, the Engineer shall make a fair determination in accordance with the Contract, taking due regard to all relevant circumstances.

The Engineer shall give notice to both Parties of each agreement or determination, with supporting particulars. Each Party shall give effect to each agreement or determination unless and until revised under Clause 20 [Claims, Disputes and Arbitration].

Part iii Construction

The Contractor

4.1 Contractor's General Obligations

The Contractor shall design (to the extent specified in the Contract), execute and complete the Works in accordance with theContract and with the Engineer's instructions, and shall remedy any defects in the Works.

The Contractor shall provide the Plant and Contractor's Documents specified in the Contract, and all Contractor's Personnel, Goods, consumables and other things and services, whether of a temporary or permanent nature, required in and for this design, execution, completion and remedying of defects.

The Contractor shall be responsible for the adequacy, stability and safety of all Site operations and of all methods of construction.

Except to the extent specified in the Contract, the Contractor

i. shall be responsible for all Contractor's Documents, Temporary Works, and such design of each item of Plant andMaterials as is required for the item to be in accordance withthe Contract, and
ii. shall not otherwise be responsible for the design or specification of the Permanent Works.

The Contractor shall, whenever required by the Engineer, submit details of the arrangements and methods which the Contractor proposes to adopt for the execution of the Works.

No significant alteration to these arrangements and methods shallbe made without this having previously been notified to theEngineer.

If the Contract specifies that the Contractor shall design any part of the Permanent Works, then unless otherwise stated in the Particular Conditions:

a) the Contractor shall submit to the Engineer the Contractor's Documents for this part in accordance with the procedures specified in the Contract;
b) these Contractor's Documents shall be in accordance with the Specification and Drawings, shall be written in the languagefor communications defined in Sub-Clause 1.4 [Law and Language], and shall include additional information requiredby the Engineer to add to the Drawings for coordination ofeach Party's designs.
c) the Contractor shall be responsible for this part, and it shall, when the Works are completed, be fit for such purposes for which the part is intended as are specified in the Contract; and
d) prior to the commencement of the Tests on Completion, the Contractor shall submit to the Engineer the "as-built" documents and operation and maintenance manuals in accordance with the Specification and in sufficient detail for the Engineer to operate, maintain, dismantle, reassemble, adjustand repair this part of the Works.

Such parts shall not be considered to be completed for the purposes of taking-over under Sub-Clause 10.1 [Taking Over of the Worksand Sections] until these documents and manuals have been submitted to the Engineer.

4.2 Performance Security

The Contractor shall obtain (at his cost) a Performance Securityfor proper performance, in the amount and currencies stated in the Appendix to Tender.

If an amount is not stated in the Appendix to Tender, this Sub-Clause shall not apply.

The Contractor shall deliver the Performance Security to the Engineer within 28 days after receiving the Letter of Acceptance and shall send a copy to the Engineer.

The Performance Security shall be issued by an entity from withina country (as per 4.1 Contractor's General Obligations; 4.2 Performance Security approved by the Employer/Government andshall be in the form annexed to the Particular Conditions of the Contract or in another form as approved by the Employer.

The Contractor shall ensure that the Performance Security is valid and enforceable until the Contractor has executed and completed the Works and remedied any defects.

If the terms of the Performance Security specify its expiry date, and the Contractor has not become entitled to receive the Performance Certificate by the date 28 days prior to the expiry date, the Contractor shall extend the validity of the Performance Security until the Works have been completed and any defects have been remedied.

The Engineer shall not make a claim under the Performance Security, except for amounts to which the Engineer is entitled under the Contract in the event of:

a. failure by the Contractor to extend the validity of thePerformance Security as described in the preceding paragraph,in which event the Engineer may claim the full amount of the Performance Security,
b. failure by the Contractor to pay the Engineer an amount due,as

either agreed by the Contractor or determined under Sub- Clause 2.5 [Engineer's Claims] or Clause 20 [Claims, Disputes and Arbitration], within 42 days after this agreementor determination,

c. failure by the Contractor to remedy a default within 42 days after receiving the Engineer's notice requiring the default to be remedied, or

d. circumstances which entitle the Engineer to termination under Sub Clause 15.2 [Termination by Engineer], irrespective of whether notice of termination has been given. The Engineer shall indemnify and hold the Contractor harmless against and from all damages, losses and expenses (including legal fees and expenses) resulting from a claim under the Performance Security to the extent to which the Engineer was not entitled to make the claim.

The Engineer shall return the Performance Security to the Contractor within 21 days after receiving a copy of the Performance Certificate.

4.3 Contractor's Representative

The Contractor shall appoint the Contractor's Representative and shall give him all authority necessary to act on the Contractor's behalf under the Contract.

Unless the Contractor's Representative is named in the Contract, the Contractor shall, prior to the Commencement Date, submit to the Engineer for consent the name and particulars of the person the Contractor proposes to appoint as Contractor's Representative.

If consent is withheld or subsequently revoked, or if the appointed person fails to act as Contractor's Representative, the Contractor shall similarly submit the name and particulars of another suitableperson for such appointment.

The Contractor shall not, without the prior consent of the Engineer, revoke the appointment of the Contractor's Representative or appoint a replacement.

The whole time of the Contractor's Representative shall be givento directing the Contractor's performance of the Contract.

Ifthe Contractor's Representative is to be temporarily absent fromthe Site during the execution of the Works, a suitable replacementperson shall be appointed, subject to the Engineer's prior consent,and the Engineer shall be notified accordingly.

The Contractor's Representative shall, on behalf of the Contractor, receive instructions under Sub-Clause 3.3 [Instructions of the Engineer]

The Contractor's Representative may delegate any powers, functions and authority to any competent person, and may at any time revoke the delegation. Any delegation or revocation shall nottake effect until the Engineer has received prior notice signed by the Contractor's Representative, naming the person and specifyingthe powers, functions and authority being delegated or revoked. The Contractor's Representative and all these persons shall be fluent in the language for communications defined in Sub-Clause 1.4 [Law and Language].

4.4 Subcontractors

The Contractor shall not subcontract the whole of the Works. The Contractor shall be responsible for the acts or defaults of any Subcontractor, his agents or employees, as if they were the acts or defaults of the Contractor.

Unless otherwise stated in the Particular Conditions:

a) the Contractor shall not be required to obtain consent tosuppliers of Materials, or to a subcontract for which the Subcontractor is named in the Contract.

b) the prior consent of the Engineer shall be obtained to other proposed Subcontractors.

c) the Contractor shall give the Engineer not less than 28 days' notice of the intended date of the commencement of each Subcontractor's work, and of the commencement of such work on the Site; and

d) each subcontract shall include provisions which would entitlethe Employer to require the subcontract to be assigned to the Engineer under Sub-Clause 4.5 [Assignment of Benefit of Subcontract] (if or when applicable) or in the event of termination under Sub-Clause 15.2 [Termination by Engineer]

4.5 Assignment of Benefit of Subcontract

If a Subcontractor's obligations extend beyond the expiry date ofthe relevant Defects Notification Period and the Engineer, prior tothis date, instructs the Contractor to assign the benefit of such obligations to the Engineer, then the Contractor shall do so.Unless otherwise stated in the assignment, the Contractor shall have no liability to the Engineer for the work carried out by the Subcontractor after the assignment takes effect.

4.6 Co-operation

The Contractor shall, as specified in the Contract or as instructedby the Engineer, allow appropriate opportunities for carrying out work to:

a. the Engineer's Personnel,

b. any other contractors employed by the Engineer, and

c. the personnel of any legally constituted public authorities, who may be employed in the execution on or near the Site of any work not included in the Contract.

Any such instruction shall constitute a Variation if and to the extent that it causes the Contractor to incur Unforeseeable Cost.

Services for these personnel and other contractors may include the use of Contractor's Equipment, Temporary Works or access arrangements which are the responsibility of the Contractor.

If, under the Contract, the Engineer is required to give to the Contractor possession of any foundation, structure, plant or meansof access in accordance with Contractor's Documents, the Contractor shall submit such documents to the Engineer in the time and manner stated in the Specification.

4.7 Setting Out

The Contractor shall set out the Works in relation to original points, lines and levels of reference specified in the Contract or notified by the Engineer.

The Contractor shall be responsible for the correct positioning ofall parts of the Works, and shall rectify any error in the positions, levels, dimensions or alignment of the Works.

The Engineer shall be responsible for any errors in these specifiedor notified items of reference, but the Contractor shall use reasonable efforts to verify their accuracy before they are used.

If the Contractor suffers delay and/or incurs Cost from executing work which was necessitated by an error in these items of reference, and an experienced contractor could not reasonably have discovered such error and avoided this delay and/or Cost, the Contractor shall give notice to the Engineer and shall be entitled subject to Sub Clause 20.1 [Contractor's Claims] to:

a) an extension of time for any such delay, if completion is or will be delayed, under Sub-Clause 8.4 [Extension of Time for Completion], and

b) payment of any such Cost-plus reasonable profit, which shallbe included in the Contract Price. After receiving this notice, the Engineer shall proceed in accordance with Sub Clause 3.5 [Determinations] to agree or determine.

i. whether and (if so) to what extent the error could notreasonably have been discovered, and

ii. the matters described in sub-paragraphs (a) and (b) above are related to this extent.

4.8 Safety Procedures

The Contractor shall:

a) comply with all applicable safety regulations,

b) take care for the safety of all persons entitled to be on the Site,

c) use reasonable efforts to keep the Site and Works clear ofunnecessary obstruction to avoid danger to these persons,

d) provide fencing, lighting, guarding and watching of the Works until completion and taking over under Clause 10 [Employer's Taking Over], and

e) provide any Temporary Works (including roadways, footways, guards and fences) which may be necessary, because of the execution of the Works, for the use and protection of the public and of owners and occupiers of adjacent land.

4.9 Quality Assurance

The Contractor shall institute a quality assurance system to demonstrate compliance with the requirements of the Contract.

The system shall be in accordance with the details stated in the Contract.

The Engineer shall be entitled to audit any aspect of the system.

Details of all procedures and compliance documents shall be submitted to the Engineer for information before each design and execution stage is commenced.

When any document of a technical nature is issued to the Engineer, evidence of the prior approval by the Contractor himself shall be apparent on the document itself.

Compliance with the quality assurance system shall not relieve the Contractor of any of his duties, obligations or responsibilities under the Contract.

4.10 Site Data

The Engineer shall have made available to the Contractor for his information, prior to the Base Date, all relevant data in the Engineer's possession on sub-surface and hydrological conditions at the Site, including environmental aspects.

The Employer 4.7 Setting Out 4.8 Safety Procedures 4.9 Quality Assurance 4.10 Site Data 16 Conditions of Contract forConstruction shall similarly make available to the Contractor all such data which come into the Employer's possession after the Base Date.

The Contractor shall be responsible for interpreting all such data.

To the extent which was practicable (taking account of cost and time), the Contractor shall be deemed to have obtained all necessary information as to risks, contingencies and other circumstances which may influence or affect the Tender or Works.

To the same extent, the Contractor shall be deemed to haveinspected and examined the Site, its surroundings, the above data and other available information, and to have been satisfied beforesubmitting the Tender as to all relevant matters, including (without limitation):

a) the form and nature of the Site, including sub-surface conditions,
b) the hydrological and climatic conditions,
c) the extent and nature of the work and Goods necessary for the execution and completion of the Works and the remedying of any defects,
d) the Laws, procedures and labor practices of the Country, and
e) the Contractor's requirements for access, accommodation, facilities, personnel, power, transport, water and other services.

4.11 Sufficiency of the Accepted Contract Amount

The Contractor shall be deemed to:

a) have satisfied himself as to the correctness and sufficiency ofthe Accepted Contract Amount, and
b) have based the Accepted Contract Amount on the data, interpretations, necessary information, inspections, examinations and satisfaction as to all relevant matters referred to in Sub-Clause 4.10 [Site Data].

Unless otherwise stated in the Contract, the Accepted Contract Amount covers all the Contractor's obligations under the Contract (including those under Provisional Sums, if any) and all things necessary for the proper execution and completion of the Works and the remedying of any defects.

4.12 Unforeseeable Physical Conditions

In this Sub-Clause, "physical conditions" means natural physical conditions and manmade and other physical obstructions and pollutants, which the Contractor encounters at the Site when executing the Works, including sub-surface and hydrological conditions but excluding climatic conditions.

If the Contractor encounters adverse physical conditions which he considers to have been Unforeseeable, the Contractor shall give notice to the Engineer as soon as practicable.

This notice shall describe the physical conditions, so that they canbe inspected by the Engineer, and shall set out the reasons why the Contractor considers them to be Unforeseeable.

The Contractor shall continue executing the Works, using such proper and reasonable measures as are appropriate for the physical conditions, and shall comply with any instructions which the Engineer may give.

If an instruction constitutes a Variation, Clause 13 [Variations and Adjustments] shall apply.

If and to the extent that the Contractor encounters physical conditions which are Unforeseeable, gives such a notice, and suffersdelay and/or incurs Cost due to these conditions, the Contractor shall be entitled subject to Sub-Clause 20.1 [Contractor's Claims] to:

a. an extension of time for any such delay, if completion is or will be delayed, under Sub-Clause 8.4 [Extension of Time for Completion], and
b. payment of any such Cost, which shall be included in the Contract Price. 4.11 Sufficiency of the Accepted Contract Amount 4.12 Unforeseeable Physical Conditions General Conditions 17 After receiving such notice and inspecting and/or investigating these physical conditions, the Engineer shall proceed in accordance with

Sub-Clause 3.5 [Determinations] to agree or determine

i. whether and (if so) to what extent these physical conditions were Unforeseeable, and

ii. the matters described in subparagraphs (a) and (b) above related to this extent. However, before additional Cost is finally agreed or determined under subparagraph (ii), theEngineer may also review whether other physical conditions in similar parts of the Works (if any) were more favorable than could reasonably have been foreseen when the Contractor submitted the Tender.

If and to the extent that these more favorable conditions were encountered, the Engineer may proceed in accordance with Sub-Clause 3.5 [Determinations] to agree or determine the reductions in Cost which were due to these conditions, which may be included (as deductions) in the Contract Price and Payment Certificates.

However, the net effect of all adjustments under sub-paragraph (b)and all these reductions, for all the physical conditions encountered in similar parts of the Works, shall not result in a net reduction inthe Contract Price.

The Engineer may take account of any evidence of the physical conditions foreseen by the Contractor when submitting the Tender, which may be made available by the Contractor, but shall not be bound by any such evidence.

4.13 Rights of Way and Facilities

The Contractor shall bear all costs and charges for special and/or temporary rights-of-way which he may require, including those for access to the Site.

The Contractor shall also obtain, at his risk and cost, any additional facilities outside the Site which he may require for the purposes of the Works.

4.14 Avoidance of Interference

The Contractor shall not interfere unnecessarily or improperly with:

a) the convenience of the public, or

b) the access to and use and occupation of all roads and footpaths, irrespective of whether they are public or in the possession ofthe Engineer or of others.

The Contractor shall indemnify and hold the Engineer harmless against and from all damages, losses and expenses (including legal fees and expenses) resulting from any such unnecessary or improper interference.

4.15 Access Route

The Contractor shall be deemed to have been satisfied as to the suitability and availability of access routes to the Site.

The Contractor shall use reasonable efforts to prevent any road or bridge from being damaged by the Contractor's traffic or by the Contractor's Personnel.

These efforts shall include the proper use of appropriate vehicles and routes. Except as otherwise stated in these Conditions:

a) the Contractor shall (as between the Parties) be responsible for any maintenance which may be required for his use of access routes.

b) the Contractor shall provide all necessary signs or directions along access routes and shall obtain any permission which may be required from the relevant authorities for his use of routes, signs and directions.

c) the Engineer shall not be responsible for any claims which may arise from the use or otherwise of any access route.

d) the Engineer does not guarantee the suitability or availabilityof particular access routes, and

e) Costs due to non-suitability or non-availability, for the use required by the Contractor, of access routes shall be borne bythe Contractor.

4.16 Transport of Goods

Unless otherwise stated in the Particular Conditions:

a. the Contractor shall give the Engineer not less than 21 days' notice of the date on which any Plant or a major item of other Goods will be delivered to the Site.

b. the Contractor shall be responsible for packing, loading, transporting, receiving, unloading, storing and protecting all Goods and other things required for the Works; and

c. the Contractor shall indemnify and hold the Engineer harmless against and from all damages, losses and expenses (including legal fees and expenses) resulting from the transport of Goods and shall negotiate and pay all claims arising from their transport.

4.17 Contractor's Equipment

The Contractor shall be responsible for all Contractor's Equipment. When brought on to the Site, Contractor's Equipment shall be deemed to be exclusively intended for the execution of the Works.

The Contractor shall not remove from the Site any major items of Contractor's Equipment without the consent of the Engineer.

However, consent shall not be required for vehicles transporting Goods or Contractor's Personnel off Site.

4.18 Protection of the Environment

The Contractor shall take all reasonable steps to protect the environment (both on and off the Site) and to limit damage and nuisance to people and property resulting from pollution, noise and other results of his operations.

The Contractor shall ensure that emissions, surface discharges and effluent from the Contractor's activities shall not exceed the values indicated in the Specification and shall not exceed the values prescribed by applicable Laws.

4.19 Electricity, Water and Gas

The Contractor shall, except as stated below, be responsible for the provision of all power, water and other services he may require.

The Contractor shall be entitled to use for the purposes of theWorks such supplies of electricity, water, gas and other services as may be available on the Site and of which details and prices are given in the Specification.

The Contractor shall, at his risk and cost, provide any apparatus necessary for his use of these services and for measuring the quantities consumed.

The quantities consumed and the amounts due (at these prices) for such services shall be agreed or determined by the Engineer in accordance with Sub-Clause 2.5 [Engineer's Claims] and Sub- Clause 3.5 [Determinations].

The Contractor shall pay these amounts to the Engineer

4.20 Engineer's Equipment and Free-Issue Material

The Engineer shall make the Engineer's Equipment (if any) available for the use of the Contractor in the execution of the Works in accordance with the details, arrangements and prices stated in the Specification.

Unless otherwise stated in the Specification:

a) the Employer shall be responsible for the Employer's Equipment, except that 4.16 Transport of Goods 4.17 Contractor's Equipment 4.18 Protection of the Environment 4.19 Electricity, Water and Gas 4.20 Employer's Equipment and Free-Issue Material General Conditions 19

b) the Contractor shall be responsible for each item of Employer's Equipment whilst any of the Contractor's Personnel is operating it, driving it, directing it or in possession or controlof it.

The appropriate quantities and the amounts due (at such stated prices) for the use of Engineer's Equipment shall be agreed or determined by the Engineer in accordance with Sub-Clause 2.5 [Engineer's Claims] and Sub-Clause 3.5

[Determinations].

The Contractor shall pay these amounts to the Engineer.

The Engineer shall supply, free of charge, the "free-issue materials"(if any) in accordance with the details stated in the Specification.

The Engineer shall, at his risk and cost, provide these materials atthe time and place specified in the Contract.

The Contractor shall then visually inspect them and shall promptly give notice to the Engineer of any shortage, defect or default in these materials.

Unless otherwise agreed by both Parties, the Engineer shall immediately rectify the notified shortage, defect or default. Afterthis visual inspection, the free-issue materials shall come under the care, custody and control of the

Contractor.

The Contractor's obligations of inspection, care, custody andcontrol shall not relieve the Engineer of liability for any shortage,defect or default not apparent from a visual inspection.

4.21 Progress Reports

Unless otherwise stated in the Particular Conditions, monthly progress reports shall be prepared by the Contractor and submitted to the Engineer in six copies.

The first report shall cover the period up to the end of the first calendar month following the Commencement Date.

Reports shall be submitted monthly thereafter, each within 7 days after the last day of the period to which it relates.

Reporting shall continue until the Contractor has completed all work which is known to be outstanding at the completion date stated in the Taking-Over Certificate for the Works.

Each report shall include:

a) charts and detailed descriptions of progress, including each stage of design (if any), Contractor's Documents, procurement, manufacture, delivery to Site, construction, erection and testing; and including these stages for work by each nominated Sub-contractor (as defined in Clause 5 [Nominated Sub-contractors]),

b) photographs showing the status of manufacture and of progress on the Site;

c) for the manufacture of each main item of Plant and Materials, the name of the manufacturer, manufacture location, percentage progress, and the actual or expected dates of: (I) commencement of manufacture, (ii) Contractor's inspections, (iii) tests, and (iv) shipment and arrival at the Site;

d) the details described in Sub-Clause 6.10 [Records of Contractor's Personnel and Equipment];

e) copies of quality assurance documents, test results andcertificates of Materials;

f) list of notices given under Sub-Clause 2.5 [Employer's Claims] and notices given under Sub-Clause 20.1 [Contractor's Claims];

g) safety statistics, including details of any hazardous incidents and activities relating to environmental aspects and public relations; and 4.21 Progress Reports 20 Conditions of Contract for Construction

h) comparisons of actual and planned progress, with details ofany events or circumstances which may jeopardize the completion in accordance with the Contract, and the measures being (or to be) adopted to overcome delays.

4.22 Security of the Site

Unless otherwise stated in the Particular Conditions:

a) the Contractor shall be responsible for keeping unauthorized persons off the Site, and

b) authorized persons shall be limited to the Contractor's Personnel and the Engineer's Personnel; and to any other personnel notified to the Contractor, by the Engineer or the Engineer, as authorized personnel of the Engineer's other contractors on the Site.

4.23 Contractor's Operations on Site

The Contractor shall confine his operations to the Site, and to any additional areas which may be obtained by the Contractor and agreed by the Engineer as working areas.

The Contractor shall take all necessary precautions to keep

Contractor's Equipment and Contractor's Personnel within the Site and these additional areas, and to keep them off adjacent land.

During the execution of the Works, the Contractor shall keep the Site free from all unnecessary obstruction and shall store or dispose of any Contractor's Equipment or surplus materials.

The Contractor shall clear away and remove from the Site any wreckage, rubbish and Temporary Works which are no longer required.

Upon the issue of a Taking-Over Certificate, the Contractor shall clear away and remove, from that part of the Site and Works to which the Taking-Over Certificate refers, all Contractor'sEquipment, surplus material, wreckage, rubbish and Temporary Works.

The Contractor shall leave that part of the Site and the Works in a clean and safe condition.

However, the Contractor may retain on Site, during the Defects Notification Period, such Goods as are required for the Contractorto fulfil obligations under the Contract.

4.24 Fossils

All fossils, coins, articles of value or antiquity, and structures and other remains or items of geological or archaeological interest found on the Site shall be placed under the care and authority of the Employer.

The Contractor shall take reasonable precautions to prevent Contractor's Personnel or other persons from removing or damaging any of these findings.

The Contractor shall, upon discovery of any such finding, promptly give notice to the Engineer, who shall issue instructions for dealing with it.

If the Contractor suffers delay and/or incurs Cost from complying with the instructions, the Contractor shall give a further notice to the Engineer and shall be entitled subject to Sub-Clause 20.1 [Contractor's Claims] to:

a) an extension of time for any such delay, if completion is or will be delayed, under Sub-Clause 8.4 [Extension of Time for Completion], and

b) payment of any such Cost, which shall be included in the Contract Price. After receiving this further notice, the Engineer shall proceed in accordance with Sub Clause 3.5 [Determinations] to agree or determine these matters.

Nominated Subcontractors

5.1 Definition of "nominated Subcontractor"

In the Contract, "nominated Subcontractor" means a Subcontractor:

a. who is stated in the Contract as being a nominated Subcontractor, or

b. whom the Engineer, under Clause 13 [Variations and Adjustments], instructs the Contractor to employ as a Subcontractor.

5.2 Objection to Nomination

The Contractor shall not be under any obligation to employ a nominated Subcontractor against whom the Contractor raises reasonable objection by notice to the Engineer as soon as practicable, with supporting particulars.

An objection shall be deemed reasonable if it arises from (among other things) any of the following matters, unless the Employer agrees to indemnify the Contractor against and from the consequences of the matter:

a) there are reasons to believe that the Subcontractor does not have sufficient competence, resources or financial strength.

b) the subcontract does not specify that the nominated Subcontractor shall indemnify the Contractor against and from any negligence or misuse of Goods by the nominated Subcontractor, his agents and employees; or

c) the subcontract does not specify that, for the subcontracted work (including design, if any), the nominated Subcontractorshall:

i. undertake to the Contractor such obligations and liabilities as will enable the Contractor to discharge his obligations and liabilities under the Contract, and

ii. indemnify the Contractor against and from all obligations and liabilities arising under or in connection with the Contract andfrom the consequences of any failure by the Subcontractor toperform these obligations or to fulfil these liabilities.

5.3 Payments to nominated Subcontractors

The Contractor shall pay to the nominated Subcontractor the amounts which the Engineer certifies to be due in accordance withthe subcontract. These amounts plus other charges shall be included in the Contract Price in accordance with subparagraph (b)of Sub-Clause 13.5 [Provisional Sums], except as stated in Sub Clause 5.4 [Evidence of Payments]

In case Subcontractors are not nominated, but working with contractors and not getting payment, and complaints reaching directly or indirectly, Engineers have the responsibility to create ateam to judge the complaint and give recommendation for payment to the Authority and the same payment should be deducted from the Contractor's payment.

In case Subcontractors are not nominated but working with contractors as Labor and complaints reaching directly or indirectly, Engineers have responsibility to judge the case of genuinely and make recommendation for the payment.

5.4 Evidence of Payments

Before issuing a Payment Certificate, which includes an amount payable to a nominated Subcontractor, the Engineer may request the Contractor to supply reasonable evidence that the nominated Subcontractor has received all amounts due in accordance with previous Payment Certificates, less applicable deductions for retention or otherwise.

Unless the Contractor:

a) submits this reasonable evidence to the Engineer, or

b) (I) satisfies the Engineer in writing that the Contractor is reasonably entitled to withhold refuse to pay these amounts, and (ii) submits to the Engineer reasonable evidence that the nominated Subcontractor has been notified of the Contractor's entitlement then the Engineer may (at his sole discretion) pay, direct to the nominated Subcontractor, part or all of such amounts previously certified (less applicable deductions) as are due to the nominated Subcontractor and for which the Contractor has failed to submit the evidence described in sub-paragraphs (a) or (b) above.

The Contractor shall then repay, to the Engineer, the amount which the nominated Subcontractor was directly paid by theEngineer.

Staff and Labor

6.1 Engagement of Staff and Labor

Except as otherwise stated in the Specification, the Contractor shall make arrangements for the engagement of all staff and labor, local or otherwise, and for their payment, housing, feeding and transport.

6.2 Rates of Wages and Conditions of Labor

The Contractor shall pay rates of wages, and observe conditions of labor, which are not lower than those established for the trade or industry where the work is carried out.

If no established rates or conditions are applicable, the Contractor shall pay rates of wages and observe conditions which are not lower than the general level of wages and conditions observed locally by Engineer's whose trade or industry is similar to that of the Contractor.

6.3 Persons in the Service of Employer, Authority and Engineer.

The Contractor shall recruit in proper channel, or attempt torecruit, staff and labor from amongst the Employer's, Authority & Engineer's Personnel.

6.4 Labor Laws

The Contractor shall comply with all the relevant labor Laws applicable to the Contractor's Personnel, including Laws relating to their employment, health, safety, welfare, immigration and emigration, and shall allow them all their legal rights. The Contractor shall require his employees to obey all applicable Laws, including those concerning safety at work

6.5 Working Hours

No work shall be carried out on the Site on locally recognized days of rest, or outside the normal working hours stated in the Appendix to Tender, unless:

a. otherwise stated in the Contract,
b. the Engineer gives consent, or
c. the work is unavoidable, or necessary for the protection of lifeor property or for the safety of the Works, in which case the Contractor shall immediately advise the Engineer.

6.6 Facilities for Staff and Labor

Except as otherwise stated in the Engineer's Requirements, the Contractor shall provide and maintain all necessary accommodation and welfare facilities for the Contractor's Personnel. The Contractor shall also provide facilities for the Engineer's Personnel as stated in the Engineer's Requirements.

The Contractor shall not permit any of the Contractor's Personnelto maintain any temporary or permanent living quarters within the structures forming part of the Permanent Works.

6.7 Health and Safety

The Contractor shall at all times take all reasonable precautions to maintain the health and safety of the Contractor's Personnel.

In collaboration with local health authorities, the Contractor shall ensure that medical staff, first aid facilities, sick bay and ambulance service are available at all times at the Site and at any accommodation for Contractor's and Employer's Personnel, and that suitable arrangements are made for all necessary welfare and hygiene requirements and for the prevention of epidemics.

The Contractor shall appoint an accident prevention officer at the Site, responsible for maintaining safety and protection against accidents.

This person shall be qualified for this responsibility and shall have the authority to issue instructions and take protective measures to prevent accidents.

Throughout the execution of the Works, the Contractor shall provide whatever is required by this person to exercise this responsibility and authority.

The Contractor shall send, to the Engineer, details of any accident as soon as practicable after its occurrence.

The Contractor shall maintain records and make reports concerning health, safety and welfare of persons, and damage to property, as the Engineer may reasonably require.

6.8 Contractor's Superintendence

Throughout the design and execution of the Works, and as long thereafter as is necessary to fulfil the Contractor's obligations, the Contractor shall provide all necessary superintendence to plan, arrange, direct, manage, inspect and test the work.

Superintendence shall be given by a sufficient number of persons having adequate knowledge of the language for communications (defined in Sub-Clause 1.4 [Law and Language]) and of the operations to be carried out (including the methods and techniques required, the hazards likely to be encountered and methods of preventing accidents), for the satisfactory and safe execution of the Works.

6.9 Contractor's Personnel

The Contractor's Personnel shall be appropriately qualified, skilled and experienced in their respective trades or occupations.

The Engineer may require the Contractor to remove (or cause to be removed) any person employed on the Site or Works, including the Contractor's Representative if applicable, who:

a. persists in any misconduct or lack of care,
b. carries out duties incompetently or negligently,
c. fails to conform with any provisions of the Contract, or
d. persists in any conduct which is prejudicial to safety, health, or the protection of the environment.

If appropriate, the Contractor shall then appoint (or cause to be appointed) a suitable replacement person.

6.10 Records of Contractor's Personnel and Equipment.

The Contractor shall submit, to the Engineer, details showing the number of each class of Contractor's Personnel and of each type of Contractor's Equipment on the Site.

Details shall be submitted each calendar month, in a form approved by the Engineer, until the Contractor has completed all work which is known to be outstanding at the completion date stated in the Taking-Over Certificate for the Works.

6.11 Disorderly Conduct

The Contractor shall at all times take all reasonable precautions to prevent any unlawful, riotous or disorderly conduct by or amongst the Contractor's Personnel, and to preserve peace and protection of persons and property on and near the Site.

Plant, Materials and Workmanship

7.1 Manner of Execution

Contractor has to prepare the presentation for approval from ENGINEER for any Plant, Materials and Workmanship and it should be discussed with the Engineer.

The Contractor shall carry out the manufacture of Plant, the production and manufacture of Materials, and all other executionof the Works:

a) in the manner (if any) specified in the Contract,

b) in a proper workmanlike and careful manner, in accordance with recognized good practice, and

c) with properly equipped facilities and non-hazardous Materials, except as otherwise specified in the Contract.

7.2 Samples

The Contractor shall submit the following samples of Materials, and relevant information, to the Engineer for consent prior to using the Materials in or for the Works:

a) manufacturer's standard samples of Materials and samplesspecified in the Contract, all at the Contractor's cost, and

b) additional samples instructed by the Engineer as a Variation.

Each sample shall be labelled as to origin and intended use in the Works.

7.3 Inspection

The Engineer's Personnel shall at all reasonable times:

a. have full access to all parts of the Site and to all places from which natural Materials are being obtained, and

b. during production, manufacture and construction (at the Site and elsewhere), be entitled to examine, inspect, measure and test the materials and workmanship, and to check the progress of manufacture of Plant and production and manufacture of Materials.

The Contractor shall give the Employer's Personnel full opportunity to carry out these activities, including providing access, facilities, permissions and safety equipment. No such activity shall relieve the Contractor from any obligation or responsibility.

The Contractor shall give notice to the Engineer whenever any work is ready and before it is covered up, put out of sight, or packaged for storage or transport.

The Engineer shall then either carry out the examination, inspection, measurement or testing without unreasonable delay, orpromptly give notice to the Contractor that the Engineer does notrequire to do so.

If the Contractor fails to give the notice, he shall, if and when required by the Engineer, uncover the work and thereafter reinstate and make good, all at the Contractor's cost.

7.4 Testing

This Sub-Clause shall apply to all tests specified in the Contract, other than the Tests after Completion (if any).

The Contractor shall provide all apparatus, assistance, documents and other information, electricity, equipment, fuel, consumables,instruments, labor, materials, and suitably qualified and experienced staff, as are necessary to carry out the specified tests efficiently.

The Contractor shall agree, with the Engineer, regarding the timeand place for the specified testing of any Plant, Materials and other parts of the works.

7 General Conditions 25 The Engineer tests or Clause 13 [Variations

and Adjustments], vary the location or details ofspecified tests, or instruct the Contractor to carry out additional tests.

If these varied or additional tests show that the tested Plant, Materials or workmanship is not in accordance with the Contract,the cost of carrying out this Variation shall be borne by the Contractor, notwithstanding other provisions of the Contract.

The Engineer shall give the Contractor not less than 24 hours' notice of the Engineer's intention to attend the tests.

If the Engineer does not attend at the time and place agreed, the Contractor may proceed with the tests, unless otherwise instructedby the Engineer, and the tests shall then be deemed to have been made in the Engineer's presence.

If the Contractor suffers delay and/or incurs Cost from complying with these instructions or as a result of a delay for which the Employer is responsible, the Contractor shall give notice to the Engineer and shall be entitled subject to Sub Clause 20.1

[Contractor's Claims] to:

a) an extension of time for any such delay, if completion is or will be delayed, under Sub-Clause 8.4 [Extension of Time for Completion], and

b) payment of any such Cost-plus reasonable profit, which shall be included in the Contract Price.

After receiving this notice, the Engineer shall proceed in accordance with Sub Clause 3.5 [Determinations] to agree or determine these matters.

The Contractor shall promptly forward to the Engineer duly certified reports of the tests.

When the specified tests have been passed, the Engineer shall endorse the Contractor's test certificate, or issue a certificate to him, to that effect. If the Engineer has not attended the tests, he shall be deemed to have accepted the readings as accurate.

7.5 Rejection

If, as a result of an examination, inspection, measurement or testing, any Plant, Materials or workmanship is found to be defective or otherwise not in accordance with the Contract, the Engineer may reject the Plant, Materials or workmanship bygiving notice to Contractor, with reasons.

The Contractor shall then promptly make good the defect and ensure that the rejected item complies with the Contract.

If the Engineer requires this Plant, Materials or workmanship to be retested, the tests shall be repeated under the same terms and conditions.

If the rejection and retesting cause the Employer to incur additional costs, the Contractor shall subject to Sub Clause 2.5 [Employer's Claims] pay these costs to the Employer.

7.6 Remedial Work

Notwithstanding any previous test or certification, the Engineer may instruct the Contractor to:

a. remove from the Site and replace any Plant or Materials whichis not in accordance with the Contract,

b. remove and re-execute any other work which is not in accordance with the Contract, and

c. execute any work which is urgently required for the safety of Works, whether because of an accident, unforeseeable event or otherwise.

The Contractor shall comply with the instruction within a reasonable time, which shall be the time (if any) specified in the instruction, or immediately if urgency is specified under sub- paragraph (c), If the Contractor fails to comply with the instruction,the Employer shall be entitled to employ and pay other persons tocarry out the work.

Except to the extent that the Contractor would have been entitledto payment for the work, the Contractor shall subject to Sub- Clause 2.5 [Employer's Claims] pay to the Employer all costs arising from this failure

7.7 Ownership of Plant and Materials

Each item of Plant and Materials shall, to the extent consistent with the Laws of the Country, become the property of the Employer at whichever is the earlier of the following times, free from liens and other encumbrances:

a) when it is delivered to the Site.
b) when the Contractor is entitled to payment of the value of Plant and Materials under Sub-Clause 8.10 [Payment for Plant and Materials in Event of Suspension]

7.8 Royalties

Unless otherwise stated in the Specification, the Contractor shall pay all royalties, rents and other payments for:

a. natural Materials obtained from outside the Site, and
b. the disposal of material from demolitions and excavations andof other surplus material (whether natural or man-made), except to the extent that disposal areas within the Site are specified in the Contract.

Commencement, Delays and Suspension

8.1 Commencement of Works

The Engineer shall give the Contractor not less than 7 days' noticeof the Commencement Date.

Unless otherwise stated in the Particular Conditions, the Commencement Date shall be within 42 days after the Contractor receives the Letter of Acceptance.

The Contractor shall commence the execution of the Works as soon as is reasonably practicable after the Commencement Dateand shall then proceed with the Works with due expedition and without delay.8.2 Time for Completion

The Contractor shall complete the whole of the Works, and each Section (if any), within the Time for Completion for the Works or Section (as the case may be), including:

a. achieving the passing of the Tests on Completion, and
b. completing all work which is stated in the Contract as being required for the Works or Section to be considered to be completed for the purposes of taking over under Sub-Clause 10.1 [Taking Over of the Works and Sections].

8.3 Program me

The Contractor shall submit a detailed time program me to the Engineer within 28 days after receiving the notice under Sub- Clause 8.1 [Commencement of Works].

The Contractor shall also submit a revised program me whenever the previous program me is inconsistent with actual progress or with the Contractor's obligations. Each program me shall include:

7.7 Ownership of Plant and Materials 7.8 Royalties 8.1 Commencement of Works 8.2 Time for Completion 8.3 Program me Commencement,

a) the order in which the Contractor intends to carry out the Works, including the anticipated timing of each stage ofdesign (if any), Contractor's Documents, procurement, manufacture of Plant, delivery to Site, construction, erection and testing,

b) each of these stages for work by each nominated Subcontractor (as defined in Clause 5 [Nominated Subcontractors]),

c) the sequence and timing of inspections and tests specified in the Contract, and

d) a supporting report which includes: (I) a general description ofthe methods which the Contractor intends to adopt, and of themajor stages, in the execution of the Works, and (ii) details showing the Contractor's reasonable estimate of the number of each class of Contractor's Personnel and of each type of Contractor's Equipment, required on the Site for each major stage.

e) Contractor has to prepare the presentation and contractor hasto present and discuss all the issue with Engineer at Engineeroffice for 1 day or more. All approval should be at same timeafter end of presentation.

Unless the Engineer, within 21 days after receiving a program me, gives notice to the Contractor stating the extent to which it does not comply with the Contract, the Contractor shall proceed in accordance with the program me, subject to his other obligations under the Contract.

The Employer's Personnel shall be entitled to rely upon the program me when planning their activities. The Contractor shall promptly give notice to the Engineer of specific probable future events or circumstances which may adversely affect the work, increase the Contract Price or delay the execution of the Works.

The Engineer may require the Contractor to submit an estimate of the anticipated effect of the future event or circumstances, and/or a proposal under Sub-Clause 13.3 [Variation Procedure].

If, at any time, the Engineer gives notice to the Contractor that a program me fails (to the extent stated) to comply with the Contractor to be consistent with actual progress and the Contractor's statedintentions, the Contractor shall submit a revised program me to theEngineer in accordance with this Sub-Clause.

8.4 Extension of Time for Completion

(e) Contractor has to prepare the presentation and contractor has to present and discuss all the issue with Engineer at Engineer office for 1 day or more. All approval should be at same time after end of presentation.

The Contractor shall be entitled subject to Sub-Clause 20.1 [Contractor's Claims] to an extension of the Time for Completion ifand to the extent that completion for the purposes ofSub-Clause10.1 [Taking Over of the Works and Sections] is or will be delayed by any of the following causes:

a) a Variation (unless an adjustment to the Time for Completion has been agreed under Sub-Clause 13.3 [Variation Procedure])or other substantial change in the quantity of an item of work included in the Contract,
b) a cause of delay giving an entitlement to extension of timeunder

a Sub-Clause of these Conditions,

c) exceptionally adverse climatic conditions,

d) Unforeseeable shortages in the availability of personnel or Goods caused by epidemic or governmental actions, or

e) any delay, impediment or prevention caused by or attributableto the Employer, the Employer's Personnel, or the Employer's other contractors on the Site.

If the Contractor considers himself to be entitled to an extension ofthe Time for Completion, the Contractor shall give notice to the Engineer in accordance with Sub Clause 20.1 [Contractor's Claims].

When determining each extension of time under Sub-Clause 20.1,the Engineer shall review previous determinations and may increase, but shall not decrease, the total extension of time.

8.5 Delays Caused by Authorities

If the following conditions apply, namely:

a) the Contractor has diligently followed the procedures laid down by the relevant legally constituted public authorities in the Country,

b) these authorities delay or disrupt the Contractor's work, and

c) the delay or disruption was Unforeseeable, then this delay or disruption will be considered as a cause of delay under subparagraph (b) of Sub-Clause 8.4 [Extension of Time for Completion].

8.6 Rate of Progress

If, at any time:

a. actual progress is too slow to complete within the Time for Completion, and/or

b. progress has fallen (or will fall) behind the current program me under Sub Clause 8.3 [Program me], other than as a result of acause listed in Sub-Clause 8.4 [Extension of Time for Completion], then the Engineer may instruct the Contractor to submit, under Sub Clause 8.3 [Program me], a revised program me and supporting report describing the revised methods which the Contractor proposes to adopt in order to expedite progress and complete within the Time for Completion.

Unless the Engineer notifies otherwise, the Contractor shall adopt these revised methods, which may require increases in the working hours and/or in the numbers of Contractor's Personnel and/or Goods, at the risk and cost of the Contractor.

If these revised methods cause the Employer to incur additional costs, the Contractor shall subject to Sub-Clause 2.5 [Employer's Claims] pay these costs to the Employer, in addition to delay damages (if any) under Sub-Clause 8.7 below.

8.7 Delay Damages

If the Contractor fails to comply with Sub-Clause 8.2 [Time for Completion], the Contractor shall be subject to Sub-Clause 2.5 [Employer's Claims] pay delay damages to the Employer for this default.

These delay damages shall be the sum stated in the Appendix to Tender, which shall be paid for every day which shall elapse between the relevant Time for Completion and the date stated in the Taking-Over Certificate.

However, the total amount due under this Sub-Clause shall notexceed the maximum amount of delay damages (if any) stated inthe Appendix to Tender.

These delay damages shall be the only damages due from the Contractor for such default, other than in the event of terminationunder Sub-Clause 15.2 [Termination by Employer] prior to completion of the Works.

These damages shall not relieve the Contractor from his obligation to complete the Works, or from any other duties, obligations or responsibilities which he may have under the Contract.

8.8 Suspension of Work

The Engineer may at any time instruct the Contractor to suspend progress of part or all of the Works.

During such suspension, the Contractor shall protect, store and secure such part or the Works against any deterioration, loss or damage.

The Engineer may also notify the cause for the suspension. If andto the extent that the cause is notified and is the responsibility ofthe Contractor, the following Sub Clauses 8.9, 8.10 and 8.11 shall not apply.

8.9 Consequences of Suspension

If the Contractor suffers delay and/or incurs Cost from complying with the Engineer's instructions under Sub-Clause 8.8 [Suspension of Work] and/or from resuming the work, the Contractor shall give notice to the Engineer and shall be entitled subject to Sub- Clause 20.1 [Contractor's Claims] to: an extension of time for any such delay, if completion is or will be delayed, under Sub-Clause

8.4 [Extension of Time for Completion], and

a) Payment of any such Cost, which shall be included in the Contract Price.

After receiving this notice, the Engineer shall proceed in accordance with Sub Clause 3.5 [Determinations] to agree or determine these matters.

The Contractor shall not be entitled to an extension of time for, orto payment of the Cost incurred in, making good the consequencesof the Contractor's faulty design, workmanship or materials, or of the Contractor's failure to protect, store or secure in accordance with Sub-Clause 8.8 [Suspension of Work].

8.10 Payment for Plant and Materials in Event of Suspension

The Contractor shall be entitled to payment of the value (as at the date of suspension) of Plant and/or Materials which have not been delivered to Site, if:

a) the work on Plant or delivery of Plant and/or Materials has been suspended for more than 28 days, and

b) the Contractor has marked the Plant and/or Materials as the Employer's property in accordance with the Engineer's instru-ctions.

8.11 Prolonged Suspension

If the suspension under Sub-Clause 8.8 [Suspension of Work] has continued for more than 84 days, the Contractor may request the Engineer's permission to proceed.

If the Engineer does not give permission within 28 days after being requested to do so, the Contractor may, by giving notice to the Engineer, treat the suspension as an omission under Clause 13 [Variations and Adjustments] of the affected part of the Works.

If the suspension affects the whole of the Works, the Contractor may give notice of termination under Sub-Clause 16.2 [Termination by Contractor].

8.12 Resumption of Work

After the permission or instruction to proceed is given, the Contractor and the Engineer shall jointly examine the Works and the Plant and Materials affected by the suspension.

The Contractor shall make good any deterioration or defect in or loss of the Works or Plant or Materials, which has occurred during the suspension.

Tests on Completion

9.1 Contractor's Obligations

The Contractor shall carry out the Tests on Completion in accordance with this Clause and Sub-Clause 7.4 [Testing], after providing the documents in accordance with sub-paragraph (d) ofSub-Clause 4.1 [Contractor's General Obligations].

The Contractor shall give to the Engineer not less than 21 days' notice of the date after which the Contractor will be ready to carry out each of the Tests on Completion.

Unless otherwise agreed, Tests on Completion shall be carried out within 14 days after this date, on such day or days as the Engineer shall instruct.

In considering the results of the Tests on Completion, the Engineer shall make allowances for the effect of any use of the Works by the Employer on the performance or other characteristics of the Works.

As soon as the Works, or a Section, have passed any Tests on Completion, the Contractor shall submit a certified report of the results of these Tests to the Engineer.

9.2 Delayed Tests

If the Tests on Completion are being unduly delayed by the Engineer, Sub-Clause 7.4 [Testing] (fifth paragraph) and/or Sub- Clause 10.3 [Interference with Tests on Completion] shall be applicable.

If the Tests on Completion are being unduly delayed by the Contractor, the Engineer may by notice require the Contractor to carry out the Tests within 21 days after receiving the notice.

The Contractor shall carry out the Tests on such day or days within that period as the Contractor may fix and of which he shallgive notice to the Engineer.

If the Contractor fails to carry out the Tests on Completion withinthe period of 21 days, the Employer's Personnel may proceed withthe Tests at the risk and cost of the Contractor.

The Tests on Completion shall then be deemed to have been carried out in the presence of the Contractor and the results of the Tests shall be accepted as accurate.

9.3 Retesting

If the Works, or a Section, fail to pass the Tests on Completion, Sub-Clause 7.5 [Rejection] shall apply, and the Engineer or the Contractor may require the failed Tests, and Tests on Completion on any related work, to be repeated under the same terms andconditions.

9.4 Failure to Pass Tests on Completion

If the Works, or a Section, fail to pass the Tests on Completion repeated under Sub Clause 9.3 [Retesting], the Engineer shall be entitled to: order further repetition of Tests on Completion under Sub-Clause 9.3 (a):

if the failure deprives the Employer substantially for the whole benefit of the Works or Section, reject the Works or Section (as the case may be), in which event the Employer shall have the same remedies as are provided in subparagraph (c) of Sub-Clause 11.4 [Failure to Remedy Defects]; or (c) issue a Taking-Over Certificate, if the Employer so requests.

In the event of sub-paragraph (c), the Contractor shall proceed in accordance with all other obligations under the Contract, and the Contract Price shall be reduced by such amount as shall beappropriate to cover the reduced value to the Employer as a resultof this failure.

Unless the relevant reduction for this failure is stated (or its method of calculation is defined) in the Contract, the Employer may require the reduction to be (I) agreed by both Parties (in full satisfaction of this failure only) and paid before this Taking-Over Certificate is issued, or (ii) determined and paid under Sub-Clause

2.5 [Employer's Claims] and Sub-Clause 3.5 [Determinations].

Employer's Taking Over

10.1 Taking Over of the Works and Sections

Except as stated in Sub-Clause 9.4 [Failure to Pass Tests on Completion], the Works shall be taken over by the Employer when

(i) the Works have been completed in accordance with the Contract, including the matters described in Sub-Clause 8.2 [Time for Completion] and except as allowed in sub-paragraph (a) below, and

(ii) a Taking-Over Certificate for the Works has been issued, or is deemed to have been issued in accordance with this Sub-Clause.

The Contractor may apply by notice to the Engineer for a Taking-Over Certificate not earlier than 14 days before the Works will, in the Contractor's opinion, be complete and ready for taking over.

If the Works are divided into Sections, the Contractor may similarly apply for a Taking-Over Certificate for each Section.

The Engineer shall, within 28 days after receiving the Contractor's application:

a) issue the Taking-Over Certificate to the Contractor, stating the date on which the Works or Section were completed in accordance with the Contract, except for any minor outstanding work and defects which will not substantially affect the use of the Works or Section for their intended purpose (either until or whilst this work is completed, and these defects are remedied); or

b) reject the application, giving reasons and specifying the work required to be done by the Contractor to enable the Taking- Over Certificate to be issued. The Contractor shall then complete this work before issuing a further notice under this Sub-Clause.

If the Engineer fails either to issue the Taking-Over Certificate orto reject the Contractor's application within the period of 28 days,and if the Works or Section (as the case may be) are substantially in accordance with the Contract, the Taking Over Certificate shallbe deemed to have been issued on the last day of that period.

10.2 Taking Over of Parts of the Works

The Engineer may, at the sole discretion of the Employer, issue a Taking-Over Certificate for any part of the Permanent Works.

The Employer shall not use any part of the Works (other than as a temporary measure which is either specified in the Contract or agreed by both Parties) unless and until the Engineer has issued aTaking-Over Certificate for this part. However, if the Employer does use any part of the Works before the Taking-Over Certificateis issued:

a) the part which is used shall be deemed to have been taken over as from the date on which it is used,
b) the Contractor shall cease to be liable for the care of such part as from this date, when responsibility shall pass to the Employer, and
c) if requested by the Contractor, the Engineer shall issue a Taking-Over Certificate for this part.

After the Engineer has issued a Taking-Over Certificate for a partof the Works, the Contractor shall be given the earliest opportunity to take such steps as may be necessary to carry out any outstanding Tests on Completion.

The Contractor shall carry out these Tests on Completion as soonas practicable before the expiry date of the relevant Defects Notification Period.

If the Contractor incurs Cost as a result of the Employer taking over and/or using a part of the Works, other than such use as is specified in the Contract or agreed by the Contractor, the Contractor shall (I) give notice to the Engineer and (ii) be entitled subject to Sub-Clause 20.1 [Contractor's Claims] to payment of any such Cost-plus reasonable profit, which shall be included in the Contract Price.

After receiving this notice, the Engineer shall proceed in accordance with Sub-Clause 3.5 [Determinations] to agree or determine this Cost and profit.

If a Taking-Over Certificate has been issued for a part of the Works (other than a Section), the delay damages thereafter for completion of the remainder of the Works shall be reduced.

Similarly, the delay damages for the remainder of the Section (if any) in which this part is included shall also be reduced.

For any period of delay after the date stated in this Taking-Over Certificate, the proportional reduction in these delay damages shall be calculated as the proportion which the value of the part so certified bears to the value of the Works or Section (as the case may be) as a whole.

The Engineer shall proceed in accordance with Sub-Clause 3.5 [Determinations] to agree or determine these proportions. The provisions of this paragraph shall only apply to the daily rate of delay damages under Sub-Clause 8.7 [Delay Damages] and shall not affect the maximum amount of these damages.

10.3 Interference with Tests on Completion

If the Contractor is prevented, for more than 14 days, fromcarrying out the Tests on Completion by a cause for which the Employer is responsible, the Employer shall be deemed to have taken over the

Works or Section (as the case may be) on the date when the Tests on Completion would otherwise have been completed. The Engineer shall then issue a Taking-OverCertificate accordingly, and the Contractor shall carry out the Tests on Completion as soon as practicable, before the expiry dateof the Defects

Notification Period.

The Engineer shall require the Tests on Completion to be carried out by giving 14 days' notice and in accordance with the relevant provisions of the Contract.

If the Contractor suffers delay and/or incurs Cost as a result of this delay in carrying out the Tests on Completion, the Contractor shall give notice to the Engineer and shall be entitled subject to Sub-Clause 20.1 [Contractor's Claims] to:

a) an extension of time for any such delay, if completion is or will be delayed, under Sub-Clause 8.4 [Extension of Time for Completion], and
b) payment of any such Cost-plus reasonable profit, which shallbe included in the Contract Price. After receiving this notice, the Engineer shall proceed in accordance with Sub Clause 3.5 [Determinations] to agree or determine these matters.

10.4 Surfaces Requiring Reinstatement

Except as otherwise stated in a Taking-Over Certificate, a certificate for a Section or part of the Works shall not be deemed to certify completion of any ground or other surfaces requiringreinstatement.

Defects Liability

11.1 Completion of Outstanding Work and Remedying Defects:

In order that the Works and Contractor's Documents, and each Section, shall be in the condition required by the Contract (fair wear and tear excepted) by the expiry date of the relevant Defects Notification Period or as soon as practicable thereafter, the Contractor shall:

a) complete any work which is outstanding on the date stated in a Taking-Over Certificate, within such reasonable time as is instructed by the Engineer, and

b) execute all work required to remedy defects or damage, as may be notified by (or on behalf of) the Engineer on or before the expiry date of the Defects Notification Period for the Works or Section (as the case may be).

If a defect appears or damage occurs, the Contractor shall be notified accordingly, by (or on behalf of) the Engineer.

11.2 Cost of Remedying Defects

All work referred to in sub-paragraph (b) of Sub-Clause 11.1 [Completion of Outstanding Work and Remedying Defects] shallbe executed at the risk and cost of the Contractor, if and to the extent that the work is attributable to:

a. any design for which the Contractor is responsible,
b. Plant, Materials or workmanship not being in accordance withthe Contract, or
c. failure by the Contractor to comply with any other obligation.

If and to the extent that such work is attributable to any other cause, the Contractor shall be notified promptly by (or on behalf of) the Engineer, and Sub-Clause 13.3 [Variation Procedure] shall apply.

11.3 Extension of Defects Notification Period

The Engineer shall be entitled subject to Sub-Clause 2.5 [Employer's Claims] to an extension of the Defects Notification Period for the Works or a Section if and to the extent that the Works, Section or a major item of Plant (as the case may be, and after taking over) cannot be used for the purposes for which theyare intended by reason of a defect or damage. However, a Defects Notification Period shall not be extended by designed Life of the work.

If delivery and/or erection of Plant and/or Materials was suspended under Sub Clause 8.8 [Suspension of Work] or Sub- Clause 16.1 [Contractor's Entitlement to Suspend Work], the Contractor's obligations under this Clause shall not apply to any defects or damage occurring more than two years after the DefectsNotification Period for the Plant and/or Materials would otherwisehave expired.

11.4 Failure to Remedy Defects

If the Contractor fails to remedy any defect or damage within a reasonable time, a date may be fixed by (or on behalf of) the Engineer, on or by which the defect or damage is to be remedied.

The Contractor shall be given reasonable notice of this date.

If the Contractor fails to remedy the defect or damage by this notified date and this remedial work was to be executed at the costof the Contractor under Sub Clause 11.2 [Cost of Remedying Defects], the Engineer may (at his option):

a) carry out the work himself or by others, in a reasonable manner and at the Contractor's cost, but the Contractor shall have no responsibility for this work; and the Contractor shall subject to Sub-Clause 2.5 [Employer's Claims] pay to the Engineer the costs reasonably incurred by the Engineer in remedying the defect or damage.

b) require the Engineer to agree or determine a reasonable reduction in the Contract Price in accordance with Sub-Clause 3.5 [Determinations]; or

c) if the defect or damage deprives the Engineer of substantially the whole benefit of the Works or any major part of the Works, takeover the Contract as a whole, or in respect of such major part which cannot be put to the intended use.

Without prejudice to any other rights, under the Contract or otherwise, the Engineer shall then be entitled to recover all sums paid for the Works or for such part (as the case may be), plus financing costs and the cost of dismantling the same, clearing the Site and returning Plant and Materials to the Contractor.

11.5 Removal of Defective Work

If the defect or damage cannot be remedied expeditiously on the Site and the Engineer gives consent, the Contractor may remove from the Site for the purposes of repair such items of Plant as are defective or damaged.

This consent may require the Contractor to increase the amount of the Performance Security by the full replacement cost of these items, or to provide other appropriate security.

11.6 Further Tests

If the work of remedying any defect or damage may affect the performance of the Works, the Engineer may require the repetition of any of the tests described in the Contract.

The requirement shall be made by notice within 28 days after the defect or damage is remedied.

These tests shall be carried out in accordance with the terms applicable to the previous tests, except that they shall be carried out at the risk and cost of the Party liable, under Sub-Clause 11.2[Cost of Remedying Defects], for the cost of the remedial work.

11.7 Right of Access

Until the Performance Certificate has been issued, the Contractor shall have such right of access to the Works as is reasonably required in order to comply with this Clause, except as may be inconsistent with the Engineer's reasonable security restrictions.

11.8 Contractor to Search

The Contractor shall, if required by the Engineer, search for the cause of any defect, under the direction of the Engineer.

Unless the defect is to be remedied at the cost of the Contractor under Sub-Clause 11.2 [Cost of Remedying Defects], the Cost ofthe search plus reasonable profit shall be agreed or determined bythe Engineer in accordance with Sub-Clause 3.5 [Determinations]and shall be included in the Contract Price.

11.9 Performance Certificate

Performance of the Contractor's obligations shall not be considered to have been completed until the Engineer has issued the Performance Certificate to the Contractor, stating the date on which the Contractor completed his obligations under the Contract.

The Engineer shall issue the Performance Certificate within 28 days after the latest of the expiry dates of the Defects NotificationPeriods, or as soon thereafter as the Contractor has supplied all theContractor's

Documents and completed and tested all the Works, including remedying any defects.

A copy of the Performance Certificate shall be issued to the Employer.

Only the Performance Certificate shall be deemed to constitute acceptance of the Works.

11.10 Unfulfilled Obligations

After the Performance Certificate has been issued, each Party shall remain liable for the fulfilment of any obligation which remains unperformed at that time.

For the purposes of determining the nature and extent of unperformed obligations, the Contract shall be deemed to remain in force.

11.11 Clearance of Site

Upon receiving the Performance Certificate, the Contractor shall remove any remaining Contractor's Equipment, surplus material, wreckage, rubbish and Temporary Works from the Site.

If all these items have not been removed within 28 days after the Engineer receives a copy of the Performance Certificate, the Engineer may sell or otherwise dispose of any remaining items.

The Engineer shall be entitled to be paid the costs incurred in connection with, or attributable to, such sale or disposal and restoring the Site. Any balance of the moneys from the sale shall be paid to the Contractor. If these moneys are less than the Engineer's costs, the Contractor shall pay the outstanding balance to the Engineer.

Part iv Financial Covenants

Measurement and Evaluation

12.1 Works to be Measured

The Works shall be measured, and valued for payment, in accordance with this Clause.

Whenever the Engineer requires any part of the Works to be measured, reasonable notice shall be given to the Contractor's Representative, who shall:

a. promptly either attend or send another qualified representativeto assist the Engineer in making the measurement, and

b. supply any particulars requested by the Engineer.

If the Contractor fails to attend or send a representative, the measurement made by (or on behalf of) the Engineer shall be accepted as accurate.

Except as otherwise stated in the Contract, wherever any Permanent Works are to be measured from records, these shall beprepared by the Engineer.

The Contractor shall, as and when requested, attend to examine and agree the records with the Engineer, and shall sign the same when agreed.

If the Contractor does not attend, the records shall be accepted as accurate.

If the Contractor examines and disagrees with the records, and/or does not sign them as agreed, then the Contractor shall give noticeto the Engineer of the respects in which the records are asserted to be inaccurate.

After receiving this notice, the Engineer shall review the records and either confirm or vary them. If the Contractor does not so givenotice to the Engineer within 14 days after being requested to examine the records, they shall be accepted as accurate.

12.2 Method of Measurement

Except as otherwise stated in the Contract and notwithstanding local practice:

a) measurement shall be made of the net actual quantity of eachitem of the Permanent Works, and

b) the method of measurement shall be in accordance with theBill of Quantities or other applicable Schedules.

12.3 Evaluation

A full-fledged Quantity & Cost Engineer team is required to calculate the overhead cost monthly basis and daily wise item rate ProjectWise under Engineer Representative. Quantity & CostEngineer has responsibility to prepare the rate of item monthly basis should present in joint meeting of Engineer, Contractor and Authority for billing. Measurement should be taken by Engineer and payment should processed as per latest rate of the project. Ratewill vary for different different supply also. (Suppose there is construction of 4 no. bridges and in all bridges different differentsteel used, rate should be different for each bridge based on steel.)

Except as otherwise stated in the Contract, the Engineer shall proceed in accordance with Sub-Clause 3.5 [Determinations] to agree or determine the Contract Price by evaluating each item of work, applying the measurement agreed or determined in accordance with the above Sub-Clauses 12.1 and 12.2 and the appropriate rate or price for the item.

For each item of work, the appropriate rate or price for the item shall be the rate or price specified for such item in the Contract forshort term contract (Maximum 6 months) Rate is derived on monthly basis to derive the overhead Cost separately for the project and adding of profit of 10% on monthly basis or, if there isno such item, specified for similar work. However, a new rate or price shall be appropriate for an item of work if:

(a)

i. the measured quantity of the item is changed by more than 10% from the quantity of this item in the Bill of Quantities oranother Schedule,
ii. this change in quantity more than 10% needs approval of Engineer.
iii. this change in quantity directly changes the Cost per unit quantity of this item by more than 1%, and this item is not specified in the Contract as a "fixed rate item"; or

(b)

i. the work is instructed under Clause 13 [Variations and Adjustments],
ii. no rate or price is specified in the Contract for this item, and
iii. no specified rate or price is appropriate because the item of work is not of similar character, or is not executed under similar conditions, as any item in the Contract.

Each new rate or price shall be derived from any relevant rates or prices in the Contract, with reasonable adjustments to take accountof the matters described in sub-paragraph (a) and/or (b), as applicable.

If no rates or prices are relevant for the derivation of a new rate or price, it shall be derived from the reasonable Cost of executing thework, together with reasonable profit, taking account of any otherrelevant matters.

Until such time as an appropriate rate or price is agreed or determined, the Engineer shall determine a provisional rate or price for the purposes of Interim Payment Certificates.

12.4 Omissions

Whenever the omission of any work form's part (or all) of a Variation, the value of which has not been agreed, if:

a. the Contractor will incur (or has incurred) cost which, if the work had not been omitted, would have been deemed to be covered by a sum forming part of the Accepted Contract Amount.

b. the omission of the work will result (or has resulted) in this sum not forming part of the Contract Price; and

c. this cost is not deemed to be included in the evaluation of any substituted work; then the Contractor shall give notice to the Engineer accordingly, with supporting particulars. Upon receiving this notice, the Engineer shall proceed in accordance with Sub-Clause 3.5 [Determinations] to agree or determine this cost, which shall be included in the Contract Price.

Variations and Adjustments

13.1 Right to Vary

Variations may be initiated by the Engineer at any time prior to issuing the Taking-Over Certificate for the Works, either by an instruction or by a request for the Contractor to submit a proposal.

The Contractor shall execute and be bound by each Variation, unless the Contractor promptly gives notice to the Engineer stating (with supporting particulars) that the Contractor cannot readily obtain the Goods required for the Variation.

Upon receiving this notice, the Engineer shall cancel, confirm or vary the instruction.

Each Variation may include:

a. changes to the quantities of any item of work included in the Contract (however, such changes do not necessarily constitutea Variation),

b. changes to the quality and other characteristics of any item ofwork,

c. changes to the levels, positions and/or dimensions of any partof the Works,

d. omission of any work unless it is to be carried out by others,

e. any additional work, Plant, Materials or services necessary forthe Permanent Works, including any associated Tests on Completion, boreholes and other testing and exploratory work, or

f. changes to the sequence or timing of the execution of the Works.

The Contractor shall not make any alteration and/or modificationof the Permanent Works, unless and until the Engineer instructs or approves a Variation.

13.2 Value Engineering

The Contractor may, at any time, submit to the Engineer a written proposal which (in the Contractor's opinion) will, if adopted,

i. accelerate completion,
ii. reduce the cost to the Employer of executing, maintaining or operating the Works,
iii. improve the efficiency or value to the Employer ofthe completed Works, or
iv. otherwise, be of benefit to the Employer. The proposal shall be prepared at the cost of the Contractor and shall include the items listed in Sub-Clause 13.3 [Variation Procedure].

If a proposal, which is approved by the Engineer, includes a change in the design of part of the Permanent Works, then unless otherwise agreed by both Parties:

a) the Contractor shall design this part,
b) sub-paragraphs (a) to(d) of Sub-Clause 4.1[Contractor's General Obligations] shall apply, and
c) if this change results in a reduction in the contract value of thispart, the Engineer shall proceed in accordance with Sub-Clause 3.5 [Determinations] to agree or determine a fee, which shall be included in the Contract Price.

This fee shall be half (50%) of the difference between the following amounts:

i. such reduction in contract value, resulting from the change, excluding adjustments under Sub-Clause 13.7 [Adjustments for Changes in Legislation] and Sub-Clause 13.8 [Adjustments for Changes in Cost], and
ii. the reduction (if any) in the value to the Employer of the varied works, taking account of any reductions in quality, anticipated life or

operational efficiencies. However, if amount:

(I) is less than amount, (ii), there shall not be a fee

13.3 Variation Procedure

If the Engineer requests a proposal, prior to instructing a Variation,the Contractor shall respond in writing as soon as practicable, either by giving reasons why he cannot comply (if this is the case) or by submitting:

a. a description of the proposed work to be performed and a program me for its execution,

b. the Contractor's proposal for any necessary modifications to the program me according to Sub-Clause 8.3 [Program me] and to the Time for Completion, and

c. the Contractor's proposal for evaluation of the Variation. The Engineer shall, as soon as practicable after receiving such proposal (under Sub Clause 13.2 [Value Engineering] orotherwise), respond with approval, disapproval or comments.

The Contractor shall not delay any work whilst awaiting a response.

Each instruction to execute a Variation, with any requirements forthe recording of Costs, shall be issued by the Engineer to the Contractor, who shall acknowledge receipt.

Each Variation shall be evaluated in accordance with Clause 12 [Measurement and Evaluation], unless the Engineer instructs or approves otherwise in accordance with this Clause.

13.4 Payment in Applicable Currencies

If the Contract provides for payment of the Contract Price in more than one currency, then whenever an adjustment is agreed, approved or determined as stated above, the amount payable ineach of the applicable currencies shall be specified.

For this purpose, reference shall be made to the actual or expected currency proportions of the Cost of the varied work, and to the proportions of various currencies specified for payment of the Contract Price.

13.5 Provisional Sums

Each Provisional Sum shall only be used, in whole or in part, in accordance with the Engineer's instructions, and the Contract Price shall be adjusted accordingly.

The total sum paid to the Contractor shall include only such amounts, for the work, supplies or services to which the Provisional Sum relates, as the Engineer shall have instructed. Foreach Provisional Sum, the Engineer may instruct:

a) work to be executed (including Plant, Materials or services tobe supplied) by the Contractor and valued under Sub-Clause13.3 [Variation Procedure]; and/or

b) Plant, Materials or services to be purchased by the Contractor,from a nominated Subcontractor (as defined in Clause 5[Nominated Subcontractors]) or otherwise; and for which there shall be included in the Contract Price:

i. the actual amounts paid (or due to be paid) by the Contractor,and

ii. a sum for overhead charges and profit, calculated as a percentage of these actual amounts by applying the relevant percentage rate (if any) stated in the appropriate Schedule.

If there is no such rate, the percentage rate stated in the Appendixto Tender shall be applied.

The Contractor shall, when required by the Engineer, produce quotations, invoices, vouchers and accounts or receipts in substantiation.

13.6 Daywork

For work of a minor or incidental nature, the Engineer may instruct that a Variation shall be executed on a daywork basis.

The work shall then be valued in accordance with the Daywork Schedule included in the Contract, and the following procedure shall apply.

If a Daywork Schedule is not included in the Contract, this Sub-Clause shall not apply.

Before ordering Goods for the work, the Contractor shall submit quotations to the Engineer.

When applying for payment, the Contractor shall submit invoices, vouchers and accounts or receipts for any Goods.

Except for any items for which the Daywork Schedule specifies that payment is not due, the Contractor shall deliver each day to the Engineer accurate statements in duplicate which shall include the following details of the resources used in executing the previous day's work:

a. the names, occupations and time of Contractor's Personnel,
b. the identification, type and time of Contractor's Equipmentand Temporary Works, and
c. the quantities and types of Plant and Materials used.

One copy of each statement will, if correct, or when agreed, be signed by the Engineer and returned to the Contractor.

The Contractor shall then submit priced statements of these resources to the Engineer, prior to their inclusion in the next Statement under Sub-Clause 14.3 [Application for Interim Payment Certificates].

13.7 Adjustments for Changes in Legislation

The Contract Price shall be adjusted to take account of anyincrease or decrease in Cost resulting from a change in the Laws ofthe Country (including the introduction of new Laws and the repeal or modification of existing Laws) or in the judicial or official governmental interpretation of such Laws, made after the Base Date, which affect the Contractor in the performance of obligations under the Contract.

If the Contractor suffers (or will suffer) delay and/or incurs (or will incur) additional Cost as a result of these changes in the Laws or insuch interpretations, made after the Base Date, the Contractor shall give notice to the Engineer and shall be entitled subject to Sub- Clause 20.1 [Contractor's Claims] to:

a. an extension of time for any such delay, if completion is or will be delayed, under Sub-Clause 8.4 [Extension of Time for Completion], and
b. payment of any such Cost, which shall be included in the Contract Price.

After receiving this notice, the Engineer shall proceed in accordance with Sub Clause 3.5 [Determinations] to agree or determine these matters.

13.8 Adjustments for Changes in Cost

In this Sub-Clause, "table of adjustment data" means the completed table of adjustment data included in the Appendix to Tender.

If there is no such table of adjustment data, this Sub-Clause shallnot apply.

If this Sub-Clause applies, the amounts payable to the Contractor shall be adjusted for rises or falls in the cost of labor, Goods andother inputs to the Works, by the addition or deduction of the amounts determined by the formulae prescribed in this Sub-Clause.

To the extent that full compensation for any rise or fall in Costs is not covered by the provisions of this or other Clauses, the Accepted Contract

Amount shall be deemed to have included amounts to cover the contingency of other rises and falls in costs.

The adjustment to be applied to the amount otherwise payable tothe Contractor, as valued in accordance with the appropriate Schedule and certified in Payment Certificates, shall be determined from formulae for each of the currencies in which the ContractPrice is payable.

Adjustment is to be applied to work valued on the basis of Cost or current prices.

In non-availability of rate in special case formulae and index should use with written permission from the Engineer.

The formulae shall be of the following general type:Pn = a + b Ln/Lo + c En/E0 + d Mn/Mo + ...

where:

"Pn" is the adjustment multiplier to be applied to the estimated contract value in the relevant currency of the work carried out in period "n",

this period being a month unless otherwise stated in the Appendixto Tender;

"a" is a fixed coefficient, stated in the relevant table of adjustment data, representing the non-adjustable portion in contractual payments;

"b", "c", "d", ... are coefficients representing the estimated proportion of each cost element related to the execution of the Works, as stated in the relevant 14.1 The Contract Price "Ln", "En", "Mn", ... are the current cost indices or reference prices forperiod "n", expressed in the relevant currency of payment, each ofwhich is applicable to the relevant tabulated cost element on the date 49 days prior to the last day of the period (to which the particular Payment Certificate relates); and

"Lo", "Eo", "Mo", ... are the base cost indices or reference prices, expressed in the relevant currency of payment, each of which is applicable to the relevant tabulated cost element on the Base Date.

The cost indices or reference prices stated in the table of adjustment data shall be used.

If their source is in doubt, it shall be determined by the Engineer.

For this purpose, reference shall be made to the values of the indices at stated dates (quoted in the fourth and fifth columns respectively of the table) for the purposes of clarification of the source, although these dates (and thus these values) may not correspond to the base cost indices.

In cases where the "currency of index" (stated in the table) is not the relevant currency of payment, each index shall be convertedinto the relevant currency of payment at the selling rate, established by the central bank of the Country, of this relevantcurrency on the above date for which the index is required to be applicable.

Until such time as each current cost index is available, the Engineer shall determine a provisional index for the issue of Interim Payment Certificates.

When a current cost index is available, the adjustment shall be recalculated accordingly.

If the Contractor fails to complete the Works within the Time for Completion, adjustment of prices thereafter shall be made using either

i. each index or price applicable on the date 49 days prior to the expiry of the Time for Completion of the Works, or

ii. the current index or price: whichever is more favorable to the Employer.

The weightings (coefficients) for each of the factors of cost statedin the table(s) of adjustment data shall only be adjusted if they have been rendered unreasonable, unbalanced or inapplicable, as a result of

Variations.

13.9 Change of Scope

i. The Engineer may, notwithstanding anything to the contrary contained in this Agreement, require the Contractor to make modifications/ alterations to the Works ("Change of Scope") as suitability.

Upon the Engineer making its intention known to the Contractor for the specific Change of Scope, be it positive or negative, the Contractor shall submit his proposal for the said Change of Scope involving additional cost or reduction in cost.

Any such Change of Scope shall be made and valued in accordance with the provisions of this Article 13.

ii. Provided that any such Change of Scope, subject to thecondition that it shall not entail any claims (e.g., Extension of Time/ Prolongation related claims), against the Employer.

iii. The Change of Scope shall mean the following:

a) change in specifications of any item of Works;

b) omission of any work from the Scope of the Project except under Clause 8.3 (iii); provided that, subject to Clause 13.5, the Engineer shall not omit any Work under this Clause in order to get it executed by any other Engineer; and / or

c) any additional Work, Plant, Materials or services which are not included in the Scope of the Work, including any associated Tests on completion of construction. 13.9 Procedure for Change of Scope

i. In the event of the Authority determining that a Change of Scope is necessary, it may direct the Engineer to issue to the Contractor a notice specifying in reasonable detail the Works and services contemplated thereunder (the "Change of Scope Notice").

The Contractor shall submit a detailed proposal as per Clause 13.9 (iii) within 15 days from the receipt of Change of Scope Notice.

ii. If the Contractor determines, not later than 90 days from the Appointed Date, that a Change of Scope to the Works is required, it shall prepare a proposal with relevant details as per Clause 13.9 (iii) at its own cost and shall submit to the Engineer to consider such Change of Scope (the "Change of Scope Request").

iii. Upon receipt of a Change of Scope Notice, the Contractor shall, with due diligence, provide to the Engineer such informationas is necessary, together with detailed proposal in support of:

a) the impact, if any, which the Change of Scope is likely to haveon the Project Completion Schedule if the works or services are required to be carried out during the Construction Period;and

b) the options for implementing the proposed Change of Scopeand the effect, if any, each such option would have on the costs and time thereof, including the following details:

i. break-up of the quantities, unit rates and cost for different items of work; and

ii. Proposed design for the Change of Scope;

iii. proposed modifications, if any, to the Work Completion Schedule of the Work. For the avoidance of doubt, the Parties expressly agree that the Contract Price shall be increased or decreased, as the case may be, on account of any such Change of Scope.

iv. The parties agree that costs and time for implementation of the proposed Change of Scope shall be determined as per the following:

a) Engineer's Representatives have responsibility to develop a system such that every month based on market rate; the Schedule of Rate should be prepared with actual overhead calculation at all Engineer's representative offices. On the basis cost should be calculated.

In Emergency cases for works where Schedule of Rates (SOR) of the concerned circle of State's Public Works Department are applicable at the Base Date are available, the same shall be applicable for determination of costs.

b) For item of Works not included in Schedule of Rates as mentioned in sub-para (a) of Clause 13.9 (iv) above, the cost of same shall be derived on the basis of established by the Engineer or in emergency MORTH Standard Data Book andthe Engineer shall determine the prevailing market rates andfor any item in respect of which MORTH Standard Data Book does not provide the requisite details, the Engineer shall determine the rate in accordance with Good Industry Practice.

c) The actual design charges shall be considered in calculation of Cost.

d) The costs of existing works or items, which are being changed/ omitted shall also be valued as per above procedure and only net cost shall be considered.

e) The reasonable time for completion of works to be taken under Change of Scope shall be determined by the Engineer on the basis of Good Industry Practice and if such time exceeds the Scheduled Completion Date, the issue of CompletionCertificate shall not be affected or delayed on account of construction of Change of Scope items/ works remaining incomplete on the date of Tests.

v. Upon consideration of the detailed proposal submitted by the Contractor under the Clause 13.9 (iii), the Engineer, within 15 (fifteen) days of receipt of such proposal, may in its sole discretion either accept such Change of Scope with modifications, if any, and initiate proceedings thereof in accordance with this Article 13 or reject the proposal and inform the Contractor of its decision and shall issue an order (the "Change of Scope Order") requiring the Contractor to proceed with the performance thereof.

For the avoidance of doubt, the Parties agree that the Contractor shall not undertake any Change of Scope withoutthe express consent of the Engineer, save and except any works necessary for meeting any Emergency, that too withverbal approval of Engineer which shall be confirmed in writing in next 3 (three) days.

In the event that the Parties are unable to agree, the Engineer may: (a) issue a Change of Scope Order requiring the Contractor to proceed with the performance thereof at the rates and conditions approved by the Engineer till the matter is resolved in accordance with Article 26; or (b) proceed in accordance with Clause 13.5.

vi. The provisions of this Agreement, insofar as they relate to Works and Tests, shall apply mutatis mutandis to the Works undertaken by the Contractor under this Article 13.

Payment for Change of Scope Payment for Change of Scope shallbe made in accordance with the payment schedule specified in the Change of Scope Order..

Restrictions on Change of Scope

i. No Change of Scope shall be executed unless the Engineer has issued the Change of Scope Order save and except any Works necessary for meeting any Emergency.

ii. The total value of all Change of Scope Orders shall not exceed10% (ten per cent) of the Contract Price. In case increased from 10% special meeting should called by Engineer for JUSTIFICATION.

iii. Notwithstanding anything to the contrary in this Article 13, ifany change is necessitated because of any default of theContractor in the performance of its obligations under this Agreement, the same shall not be deemed to be Change of Scope and shall not result in any adjustment of the Contract Price or the Project Completion Schedule.

13.10 Power of the Engineer to undertake Works

(I) In the event the Parties are unable to agree to the proposed Change of Scope Orders in accordance with Clause, the Engineer may, after giving notice to the Contractor and considering its reply thereto, award such Works or services to any person or agency on the basis of open competitive bidding.

It is also agreed that the Contractor shall provide assistance and cooperation to the person or agency who undertakes the works or services hereunder. The

Contractor shall not be responsible for rectification of any Defects, but the Contractor shall carry out maintenance of such works after completion of Defect Liability Period of work by other person or agency during the remaining period of this agreement without any extra payment. (ii) The Works undertaken in accordance with this Clause shall conform to the Specifications and Standards and shall be carried out in a manner that minimizes the disruption in operation of the Work. The provisions of this Agreement, insofar as they relate to Works and Tests, shall apply mutatis mutandis tothe Works carried out under this Clause 13.9

Contract Price and Payment

14.1 The Contract Price

1. Contract Price

(i) The Engineer shall make payments to the Contractor for the Works on the basis of the lump sum price accepted by the Engineer in consideration of the obligations specified in this Agreement foran amount of INR … … … … (IN...)

(ii) (the "Contract Price"), which shall be subject to adjustments in accordance with the provisions of this Agreement.

For the avoidance of doubt, the Parties expressly agree that the Contract Price shall not include the cost of Maintenance, which shall be paid separately in accordance with the provisions of Clause 19.7.

The Parties further agree that save and except as provided in this Agreement, the Contract Price shall be valid and effective until issue of Completion Certificate.

(iii) The Contract Price includes all duties, taxes, royalty, cess, charges, and fees that may be levied in accordance with the laws and regulations in force as on the Base Date on the Contractor's equipment, Plant, Materials and supplies acquired for the purpose of this Agreement and on the services performed under this Agreement.

Nothing in this Agreement shall relieve the Contractor from its responsibility to pay any tax including any tax that may be leviedin India on profits made by it in respect of this Agreement.

Unless otherwise stated in the Particular Conditions:

a) the Contract Price shall be agreed or determined under Sub-

Clause 12.3 [Evaluation] and be subject to adjustments in accordance with the Contract;

b) the Contractor shall pay all taxes, duties and fees required tobe paid by him under the Contract, and the Contract Priceshall not be adjusted for any of these costs except as stated in Sub-Clause 13.7 [Adjustments for Changes in Legislation];

c) any quantities which may be set out in the Bill of Quantities or other Schedule are estimated quantities and are not to be taken as the actual and correct quantities:

i. of the Works which the Contractor is required to execute, or

ii. for the purposes of Clause 12 [Measurement and Evaluation];and

d) the Contractor shall submit to the Engineer, within 28 days after the Commencement Date, a proposed breakdown of each lump sum price in the Schedules. The Engineer may take account of the breakdown when preparing Payment Certificates but shall not be bound by it.

14.2 Advance Payment

The Engineer shall make an advance payment, as an interest-free loan for mobilization, when the Contractor submits a guarantee in accordance with this Sub Clause.

The total advance payment, the number and timing of installments(if more than one), and the applicable currencies and proportions,shall be as stated in the Appendix to Tender.

Unless and until the Engineer receives this guarantee, or if the total advance payment is not stated in the Appendix to Tender, this Sub-Clause shall not apply.

The Engineer shall issue an Interim Payment Certificate for thefirst instalment after receiving a

Statement (under Sub-Clause 14.3 [Application for Interim Payment

Certificates]) and after the Engineer receives the Performance Security in accordance with Sub-Clause 4.2 [Performance Security] and

a. a guarantee in amounts and currencies equal to the advance payment.

This guarantee shall be issued by an entity and from within a country (or other jurisdiction) approved by the Engineer and shall be in the form annexed to the Particular Conditions or in another form approved by the Engineer.

The Contractor shall ensure that the guarantee is valid and enforceable until the advance payment has been repaid, but its amount may be progressively reduced by the amount repaid by theContractor as indicated in the Payment Certificates.

If the terms of the guarantee specify its expiry date, and the advance payment has not been repaid by the date 28 days prior tothe expiry date, the Contractor shall extend the validity of the guarantee until the advance payment has been repaid.

The advance payment shall be repaid through percentage deductions in Payment Certificates. Unless other percentages are stated in the Appendix to Tender:

a) deductions shall commence in the Payment Certificate in which the total of all certified interim payments (excluding the advance payment and deductions and repayments of retention) exceeds ten per cent (10%) of the Accepted Contract Amount Less Provisional Sums; and
b) deductions shall be made at the amortization rate of one quarter (25%) of the amount of each Payment Certificate (excluding the advance payment and deductions and repayments of retention) in the currencies and proportions ofthe advance payment, until such time as the advance paymenthas been repaid.

If the advance payment has not been repaid prior to the issue of the Taking-Over Certificate for the Works or prior to termination under Clause 15 [Termination by Employer], Clause 16[Suspension and Termination by Contractor] or Clause 19 [ForceMajeure] (as the case may be), the whole of the balance then outstanding shall immediately become due and payable by the Contractor to the Employer.

14.3 Application for Interim Payment Certificates

The Contractor shall submit a Statement in six copies to the Engineer after the end of each month, in a form approved by the Engineer, showing in detail the amounts to which the Contractor considers himself to be entitled, together with supporting documents which shall include the report on the progress during this month in accordance with Sub-Clause 4.21 [Progress Reports].

The Statement shall include the following items, as applicable, which shall be expressed in the various currencies in which the Contract Price is payable, in the sequence listed:

the estimated contract value of the Works executed and the Contractor's Documents produced up to the end of the month (including Variations but excluding items described in sub- paragraphs (b) to (g) below);

a. any amounts to be added and deducted for changes in legislation and changes in cost, in accordance with Sub-Clause

13.7 [Adjustments for

b. Changes in Legislation] and Sub-Clause 13.8 [Adjustments for Changes in Cost];

c. any amount to be deducted for retention, calculated by applying the percentage of retention stated in the Appendix toTender to the total of the above amounts, until the amount soretained by the Engineer reaches the limit of Retention Money(if any) stated in the Appendix to Tender;

d. any amounts to be added and deducted for the advance payment and repayments in accordance with Sub-Clause 14.2 [Advance Payment];

e. any amounts to be added and deducted for Plant and Materialsin accordance with Sub-Clause 14.5 [Plant and Materials intended for the Works];

f. any other additions or deductions which may have become due under the Contract or otherwise, including those under Clause20 [Claims, Disputes and Arbitration]; and

g. the deduction of amounts certified in all previous Payment Certificates.

14.4 Schedule of Payments

If the Contract includes a schedule of payments specifying the installments in which the Contract Price will be paid, then unless otherwise stated in this schedule:

a) the installments quoted in this schedule of payments shall be the estimated contract values for the purposes of sub- paragraph (a) of Sub-Clause 14.3 [Application for Interim Payment Certificates];

b) Sub-Clause 14.5 [Plant and Materials intended for the Works]shall not apply; and

c) if these installments are not defined by reference to the actual progress achieved in executing the Works, and if actual progress is found to be less than that on which this schedule ofpayments was based, then the Engineer may proceed in accordance with Sub-Clause 3.5 [Determinations] to agree or determine revised installments, which shall take account of the extent to which progress is less than that on which the instalment were previously based.

If the Contract does not include a schedule of payments, the Contractor shall submit non-binding estimates of the payments which he expects to become due during each quarterly period.

The first estimate shall be submitted within 42 days after the Commencement Date.

Revised estimates shall be submitted at quarterly intervals, until the Taking-Over Certificate has been issued for the Works.

14.5 Plant and Materials intended for the Works

If this Sub-Clause applies, Interim Payment Certificates shall include, under subparagraph (e) of Sub-Clause 14.3,

i. an amount for Plant and Materials which have been sent to theSite for incorporation in the Permanent Works, and

ii. a reduction when the contract value of such Plant and Materials is included as part of the Permanent Works under subparagraph (a) of Sub-Clause 14.3 [Application for InterimPayment Certificates]. If the lists referred to in sub- paragraphs (b) (I) or (c) (I) below are not included in the Appendix to Tender, this Sub-Clause shall not apply.

The Engineer shall determine and certify each addition if the following conditions are satisfied:

(a) the Contractor has:

i. kept satisfactory records (including the orders, receipts, Costsand use of Plant and Materials) which are available for inspection, and

ii. submitted a statement of the Cost of acquiring and deliveringthe Plant and Materials to the Site, supported by satisfactory evidence;

and either:

(b) the relevant Plant and Materials:

i. are those listed in the Appendix to Tender for payment when shipped,

ii. have been shipped to the Country, en route to the Site, in accordance with the Contract; and

iii. are described in a clean shipped bill of lading or other evidence of shipment, which has been submitted to the Engineer together with evidence of payment of freight and insurance, any other documents reasonably required, and a bank guarantee in a form and issued by an entity approved by the Engineer in amounts and currencies equal to the amount due under this Sub-Clause: this guarantee may be in a similar formto the form referred to in Sub-Clause 14.2 [Advance Payment]and shall be valid until the Plant and Materials are properly stored on Site and protected against loss, damage or deterioration; or (c) the relevant Plant and Materials:

a. are those listed in the Appendix to Tender for payment when delivered to the Site, and

b. have been delivered to and are properly stored on the Site, are protected against loss, damage or deterioration, and appear tobe in accordance with the Contract.

The additional amount to be certified shall be the equivalent of eighty percent of the Engineer's determination of the cost of the Plant and Materials (including delivery to Site), taking account of the documents mentioned in this Sub-Clause and of the contract value of the Plant and Materials.

The currencies for this additional amount shall be the same as those in which payment will become due when the contract value is included under sub-paragraph (a) of Sub Clause 14.3 [Applicationfor Interim Payment Certificates].

At that time, the Payment Certificate shall include the applicable reduction which shall be equivalent to, and in the same currenciesand proportions as, this additional amount for the relevant Plant and Materials.

14.6 Issue of Interim Payment Certificates

No amount will be certified or paid until the Engineer has received and approved the Performance Security.

Thereafter, the Engineer shall, within 28 days after receiving a Statement and supporting documents, issue to the Engineer an Interim Payment Certificate which shall state the amount which the Engineer fairly determines to be due, with supporting particulars.

However, prior to issuing the Taking-Over Certificate for the Works, the Engineer shall not be bound to issue an InterimPayment Certificate in an amount which would (after retention and other deductions) be less than the minimum amount of Interim Payment Certificates (if any) stated in the Appendix to Tender.

In this event, the Engineer shall give notice to the Contractor accordingly

An Interim Payment Certificate shall not be withheld for any other reason, although:

a) if anything supplied or work done by the Contractor is not in accordance with the Contract, the cost of rectification or replacement may be withheld until rectification or replacement has been completed; and/or
b) if the Contractor was or is failing to perform any work orobligation

in accordance with the Contract, and had been so notified by the Engineer, the value of this work or obligation may be withheld until the work or obligation has been performed. The Engineer may in any Payment Certificate make any correction or modification that should properly be made to any previous Payment Certificate.

A Payment Certificate shall not be deemed to indicate the Engineer's acceptance, approval, consent or satisfaction.

14.7 Payment

The Engineer shall pay to the Contractor:

a. the first instalment of the advance payment within 42 days after issuing the Letter of Acceptance or within 21 days afterreceiving the documents in accordance with Sub-Clause 4.2 [Performance Security] and Sub-Clause 14.2 [Advance Payment], whichever is later;

b. the amount certified in each Interim Payment Certificatewithin 56 days after the Engineer receives the Statement and supporting documents; and

c. the amount certified in the Final Payment Certificate within 56 days after the Engineer receives this Payment Certificate. Payment of the amount due in each currency shall be madeinto the bank account, nominated by the Contractor, in the payment country (for this currency) specified in the Contract.

14.8 Delayed Payment

If the Contractor does not receive payment in accordance with Sub-Clause 14.7 [Payment], the Contractor shall be entitled to receive financing charges compounded monthly on the amount unpaid during the period of delay.

This period shall be deemed to commence on the date for payment specified in Sub-Clause 14.7 [Payment], irrespective (in the case of its sub-paragraph (b)) of the date on which any Interim Payment Certificate is issued.

Unless otherwise stated in the Particular Conditions, these financing charges shall be calculated at the annual rate of three percentage points above the discount rate of the central bank in the country of the currency of payment and shall be paid in such currency.

The Contractor shall be entitled to this payment without formal notice or certification, and without prejudice to any other er right or remedy.

14.9 Payment of Retention Money

When the Taking-Over Certificate has been issued for the Works,the first half of the Retention Money shall be certified by the Engineer for payment to the

Contractor.

If a Taking-Over Certificate is issued for a Section or part of the Works, a proportion of the Retention Money shall be certified and paid.

This proportion shall be two-fifths (40%) of the proportion calculated by dividing the estimated contract value of the Sectionor part, by the estimated final Contract Price.

Promptly after the latest of the expiry dates of the Defects Notification Periods, the outstanding balance of the Retention Money shall be certified by the Engineer for payment to the Contractor.

If a Taking-Over Certificate was issued for a Section, a proportionof the second half of the Retention Money shall be certified andpaid promptly after the expiry date of the Defects Notification Period for the Section.

This proportion shall be two-fifths (40%) of the proportion calculated by dividing the estimated contract value of the Section by the estimated final Contract Price.

However, if any work remains to be executed under Clause 11[Defects Liability], the Engineer shall be entitled to withhold certification of the estimated cost of this work until it has been executed.

When calculating these proportions, no account shall be taken of any adjustments under Sub-Clause 13.7 [Adjustments for Changes in Legislation] and Sub Clause 1 3.8 [Adjustments for Changes inCost].

14.10 Statement at Completion

Within 84 days after receiving the Taking-Over certificate for the Works, the Contractor shall submit to the Engineer six copies of a Statement at completion with supporting documents, in accordance with Sub-Clause 14.3 [Application for Interim Payment Certificates], showing:

a. the value of all work done in accordance with the Contract upto the date stated in the Taking-Over Certificate for theWorks,

b. any further sums which the Contractor considers to be due, and

c. an estimate of any other amounts which the Contractor considers will become due to him under the Contract.

Estimated amounts shall be shown separately in this Statement at completion.

The Engineer shall then certify in accordance with Sub-Clause 14.6 [Issue of Interim Payment Certificates].

14.11 Application for Final Payment Certificate

Within 56 days after receiving the Performance Certificate, the Contractor shall submit, to the

Engineer, six copies of a draft final statement with supporting documents showing in detail in a form approved by the Engineer:

a) the value of all work done in accordance with the Contract,and
b) any further sums which the Contractor considers to be due tohim under the Contract or otherwise.

If the Engineer disagrees with or cannot verify any part of thedraft final statement, the Contractor shall submit such further information as the Engineer may reasonably require and shall make such changes in the draft as may be agreed between them.

The Contractor shall then prepare and submit to the Engineer the final statement as agreed.

This agreed statement is referred to in these Conditions as the "Final Statement".

However, if following discussions between the Engineer and the Contractor and any changes to the draft final statement which are agreed, it becomes evident that a dispute exists, the Engineer shall deliver to the Engineer (with a copy to the Contractor) an Interim Payment Certificate for the agreed parts of the draft final statement.

Thereafter, if the dispute is finally resolved under Sub-Clause 20.4 [Obtaining Dispute Adjudication Board's Decision] or Sub-Clause

20.5 [Amicable Settlement], the Contractor shall then prepare and submit to the Engineer (with a copy to the Engineer) a Final Statement.

14.12 Discharge

When submitting the Final Statement, the Contractor shall submita written discharge which confirms that the total of the Final Statement represents full and final settlement of all moneys due tothe Contractor under or in connection with the Contract.

This discharge may state that it becomes effective when the Contractor has received the Performance Security and the outstanding balance of this total, in which event the discharge shall be effective on such date.

14.13 Issue of Final Payment Certificate

Within 28 days after receiving the Final Statement and written discharge in accordance with Sub-Clause 14.11 [Application for Final Payment Certificate] and

Sub-Clause 14.12 [Discharge], the Engineer shall issue, to the Engineer, the Final Payment Certificate which shall state:

a. the amount, which is finally due, and
b. after giving credit to the Engineer for all amounts previouslypaid by the Engineer and for all sums to which the Engineer isentitled, the balance (if any) due from the Engineer to the Contractor or from the Contractor to the Engineer, as the casemay be. If the Contractor has not applied for a Final Payment Certificate in accordance with Sub-Clause 14.11 [Application for Final Payment Certificate] and Sub-Clause 14.12 [Discharge], the Engineer shall request the Contractor to do so.

If the Contractor fails to submit an application within a period of28 days, the Engineer shall issue the Final Payment Certificate for such amount as he fairly determines to be due.

14.14 Cessation of Engineer's Liability

The Engineer shall not be liable to the Contractor for any matter or thing under or in connection with the Contract or execution of the Works, except to the extent that the Contractor shall have included an amount expressly for it:

a) in the Final Statement and also
b) (except for matters or things arising after the issue of the Taking-Over Certificate for the Works) in the Statement at completion described in Sub-Clause 14.10 [Statement at Completion].

However, this Sub-Clause shall not limit the Engineer's liability under his indemnification obligations, or the Engineer's liability in any case of fraud, deliberate default or reckless misconduct by the Engineer.

14.15 Currencies of Payment

The Contract Price shall be paid in the currency or currencies named in the Appendix to Tender.

Unless otherwise stated in the Particular Conditions, if more than one currency is so named, payments shall be made as follows:

(a) if the Accepted Contract Amount was expressed in Local Currency only:

i. the proportions or amounts of the Local and Foreign Currencies, and the fixed rates of exchange to be used forcalculating the payments, shall be as stated in the Appendix to Tender, except as otherwise agreed by both Parties;

ii. payments and deductions under Sub-Clause 13.5 [Provisional Sums] and Sub-Clause 13.7 [Adjustments for Changes in Legislation] shall be made in the applicable currencies and proportions; and

iii. other payments and deductions under sub-paragraphs (a) to (d) of Sub Clause 14.3

[Application for Interim Payment Certificates] shall be made in the currencies and proportions specified in sub-paragraph (a) (I) above;

a) payment of the damages specified in the Appendix to Tendershall be made in the currencies and proportions specified in theAppendix to Tender;

b) other payments to the Engineer by the Contractor shall be made in the currency in which the sum was expended by the Engineer, or in such currency as may be agreed by both Parties;

c) if any amount payable by the Contractor to the Engineer in a particular currency exceeds the sum payable by the Engineerto the Contractor in that currency, the Engineer may recover the balance of this amount from the sums otherwise payable tothe Contractor in other currencies; and

d) if no rates of exchange are stated in the Appendix to Tender, they shall be those prevailing on the Base Date and determined by the central bank of the Country.

14.16 Price adjustment for Works for Lump Sum Work

14.16.1 The amounts payable to the Contractor for Works shall be adjusted as per rate finalized on same months in accordance with the provisions of this Clause.

14.16.2 Subject to the provisions of Clause 14.16.3(In special Case), the amounts payable to the Contractor for Works shall be adjusted in the IPC issued by the Engineer for the increase or decrease in the index cost of inputs for the works, by the additionor subtraction of the amounts determined by the formulae specified in Clause 14.16.4.

14.16.3 To the extent that any compensation or reimbursement for increase or decrease in costs to the Contractor is not covered by the provisions of this or other Clauses in this Agreement, the costsand prices payable under this Agreement shall be deemed to include the amounts required to cover the contingency of such other increase or decrease of costs and prices.

14.16.4 The Contract Price shall be adjusted for increase or decrease in rates and prices of labor, Materials, fuel and lubricants, equipment, Machinery, Plant and other Materials or inputs as finalized on end of months in special case with approval of Engineer in accordance with the principles, procedures and formulae specified below:

a. Price adjustment shall be applied on completion of the specified stage of the respective item of work in accordancewith Schedule-G;

b. Adjustment for each item of work/stage shall be madeseparately;

c. The following expressions and meanings are assigned to the value of the work done for civil and track work:

EW = Value of work done for the completion of a stage under the item earthwork;

BRIMP = Value of work done for the completion of a stage under the item Important Bridges;

BR = Value of work done for the completion of a stage under the items Major Bridges, Minor Bridges, RCC box/pipe culverts, Flyovers, RUB, and ROB in accordance with Schedule-G;

TRK = Value of work done for the completion of a stage under the item Track Works;

TUNL = Value of work done for the completion of a stage under the items Tunnel;

OEW = Value of work done for the completion of a stage under the item Other Engineering Works;

INVCIV=Value of work done for under the item inventory;

INTGTESTCIV = Value of work done for the item integrated testing and commissioning of the Railway Project.

(d) Price adjustment for change in costs of civil and track work shall be paid in accordance with the following formula:

i. VEW= 0.85 EW x [PLB x (LBi – LBo)/LBo + PC x (Ci – Co)/Co + PF x (Fi –Fo)/Fo + PMACH x (MACHi – MACHo)/MACHo + POTH x (OTHi - OTHo)/OTHo];

ii. VBRIMP = 0.85 BRIMP x [PLB x (LBi – LBo)/LBo + PC x (Ci – Co)/Co + PS x (Si – So)/So + PF x (Fi –Fo)/Fo + PMACH x (MACHi – MACHo)/MACHo + POTH x (OTHi - OTHo)/OTHo];

iii. VBR = 0.85 BR x [PLB x (LBi – LBo)/LBo + PC x (Ci – Co)/Co + PS x (Si – So)/So + PF x (Fi – Fo)/Fo + PMACH x (MACHi – MACHo)/MACHo + POTH x (OTHi - OTHo)/OTHo];

iv. VTRK = 0.85 TRK x [PLB x (LBi – LBo)/LBo + PC x (Ci – Co)/Co + PS x(Si – So)/So + PF x (Fi –Fo)/Fo + PMACH x (MACHi – MACHo)/MACHo + POTH x (OTHi – OTHo)/OTHo + PR x (Ri - Ro)/Ro];

v. VTUNL = 0.85 TUNL x [PLB x (LBi – LBo)/LBo + PC x (Ci – Co)/Co + PS x (Si – So)/So + PF x (Fi – Fo)/Fo + PMACHx (MACHi – MACHo)/MACHo + POTH x (OTHi - OTHo)/OTHo + PXLP x (XLPi – XLPo)/XLPo];

vi. VOEW = 0.85 OEW x [PLB x (LBi – LBo)/LBo + PC x (Ci – Co)/Co + PS x(Si – So)/So + PF x (Fi – Fo)/Fo + PMACH x ((MACHi – MACHo)/MACHo + POTH x (OTHi - OTHo)/OTHo];

vii. VINVCIV = 0.85 INVCIV x [PR x (Ri – Ro)/Ro + POTH x (OTHi - OTHo)/OTHo]; and

viii. VINTGTESTCIV = 0.85 INTGTESTCIV x [PLB x (LBi – LBo)/LBo = POTH x (OTHi - OTHo)/OTHo]; Where VEW = Increase or decrease in the cost of earthwork during the period under consideration due to changes in the rates for relevant components as specified in sub-paragraph (e);

VBRIMP = Increase or decrease in the cost of Important Bridges during the period under consideration due to changes in the rates for relevant components as specified in sub-paragraph (e);

VBR = Increase or decrease in the cost of Major Bridges, Minor Bridges, Flyovers, RCC box/pipe culverts ROB/RUB) during the period under consideration due to changes in the rates for relevant components as specified in sub-paragraph (e);

VTRK = Increase or decrease in the cost of track works during the period under consideration due to changes in the rates for relevant components as specified in sub-paragraph (e);

VTUNL = Increase or decrease in the cost of tunnels during the period under consideration due to changes in the rates for relevant components as specified in sub-paragraph (e);

VOEW = Increase or decrease in the cost of Other Engineering Works during the period under consideration due to changes in therates for relevant components as specified in sub-paragraph (e);

VINVCIV = Increase or decrease in the cost of inventory during the period under consideration due to changes in the rates for relevant components as specified in sub-paragraph (e);

VINTGTESTCIV = Increase or decrease in the cost of integrated testing and commissioning during the period under considerationdue to changes in the rates for relevant components as specified in subparagraph (e);

PC, PF, PLB, PMACH, POTH, PR, PS and PXLP are the

percentages of cement, fuel and lubricants, labor, Plant Machinery and tools, other materials, rails, steel/ components (includingstrands and steel cables), and explosives respectively for the relevant item as specified in sub-paragraph (e);

Co = The wholesale price index as published by the Ministry of

Commerce and Industry, Government of India (hereinafter called "WPI") for grey cement for the month of the Base Date;

Ci = The WPI for grey cement for the month which is three months prior to the month to which the IPC relates;

Fo = The official retail price of high-speed diesel (HSD) oil at the existing consumer pumps of Indian Oil Corporation ("IOC") in theState of [Karnataka] on the Base Date;

Fi = The official retail price of HSD at the existing consumer pumps of IOC in the State of [Karnataka] on the first day of the month three months prior to the month to which the IPC relates;

LBo = The consumer price index for industrial workers for the [circle **** in the State of Karnataka], published by Labor Bureau, Ministry of Labor, Government of India, (hereinafter called "CPI") for the month of the Base Date;

LBi = The CPI for the month which is three months prior to the month to which the IPC relates;

MACHo = The WPI for construction machinery for the month ofthe Base Date;

MACHi = The WPI for construction machinery for the month which is three months prior to the month to

which the IPC relates;

OTHo = The WPI for all commodities for the month of the Base Date;

OTHi = The WPI for all commodities for the month which is three months prior to the month to which the IPC relates;

Ro = The Price for Rails (60kg) published by the Bhilai Plant of the Steel Authority of India for the month of the Base Date;

Ri = The Price for Rails (60kg) published by the Bhilai Plant of the

Steel Authority of India for the month which is three months priorto the month to which the IPC relates;

So = The WPI for steel (rods) for the month of the Base Date;

Si = The WPI for steel (rods) for the month which is three months prior to the month to which the IPC relates;

XLPo = The WPI for explosives for the month of the Base Date; and

XLPi = The WPI for explosives for the month which is three months prior to the month to which the IPC relates.

(e) The following percentages shall govern the price adjustment ofthe Contract Price

Component	Earthwork	Important Bridges	Major Bridges/ fluovers/ Minor Bridges, CC Box/ Pipe Culverts/ROB/ RUB	Trackworks	Tunnels	Otherengineeringwor	Inventory	Integrated Testing and Commissioning
-1	-2	-3	-4	-5	-6	-7	-8	
Cement (PC)	***%	***%	***%	***%	***%	***%	-	-
Explosives (PXLP)	-	-	-	-	***%	-	-	-
Fuel and lubricants (PF)	***%	***%	***%	***%	***%	***%	-	-
Labour (PLB)	***%	***%	***%	***%	***%	***%	-	***%
Machinery and Plants (PMACH)	***%	***%	***%	***%	***%	***%	-	
Other Materials (POTH)	***%	***%	***%	***%	***%	***%	***%	***%
Rail (PR)	-	-	-	***%	-	-	***%	-
Steel (PS)	-	***%	***%	***%	***%	***%	-	-
Total	100%	100%	100%	100%	100%	100%	100%	100%

(a) The following expressions and meanings are assigned to the value of the work done for signaling and telecommunication works:

SIGWK = Value of signal-ling works for a stage payment of the item signal-ling works;

INVSIG = Value of inventory for signal-ling works for a stage payment of the item inventory for signal-ling works;

INTGTESTSIG = Value of integrated testing and commission for signal-ling works of the Railway Project;

COMWK= Value of telecommunication works for a stage payment of the item telecommunication works;

INVCOM = Value of inventory for telecommunication works for a stage payment of the item inventory for telecommunication works; and

INTGTESTCOM = Value of integrated testing and commission for telecommunication works of the Railway Project.

(b) Price adjustment for changes in cost of signal-ling works and telecommunication works shall be paid in accordance with the following formula:

VSIGWK = 0.85 SIGWK x [PELEX x (ELEXi – ELEXo)/ ELEXo + POFC x (OFCi – OFCo)/OFCo + PLB x (LBi –LBo)/LBo + POTH x (OTHi - OTHo)/OTHo + PIC x (ICi – ICo)/ICo];

VINVSIG = 0.85 SIGWK x [PELEX x (ELEXi – ELEXo)/ ELEXo + POTH x (OTHi - OTHo)/OTHo];

VINTGTESTSIG = 0.85 INTGTESTSIG x [PLB x (LBi – LBo)/LBo + POTH x (OTHi - OTHo)/OTHo];

VCOMWK = 0.85 COMWK x [PELEX x (ELEXi – ELEXo)/ ELEXo + POFC x (OFCi – OFCo)/OFCo + PLB x (LBi – LBo)/LBo + POTH x (OTHi - OTHo)/OTHo + PIC x (ICi – ICo)/ICo + PCEQP x (CEQPi – CEQPo)/CEQPo];

VINVCOM = 0.85 SIGWK x [PELEX x (ELEXi – ELEXo)/ELEXo + PCEQP x (CEQPi – CEQPo)/CEQPo + POTH x (OTHi - OTHo)/OTHo]; and

VINTGTESTCOM = 0.85 INTGTESTCOM x [PLB x (LBi – LBo)/LBo + POTH x (OTHi - OTHo)/OTHo].

Were

VSIGWK = Increase or decrease in the cost of signaling works during the period under consideration due to changes in the rates for relevant components as specified in sub-paragraph (h);

VINVSIG = Increase or decrease in the cost of inventory for signaling during the period under consideration due to changes inthe rates for relevant components as specified in sub-paragraph (h);

VINTGTESTSIG = Increase or decrease in the cost of integrated testing and commissioning of signaling works of the Railway Project during the period under consideration due to changes in the rates for relevant components as specified in sub-paragraph (h);

VCOMWK = Increase or decrease in the cost of communication works during the period under consideration due to changes in therates for relevant components as specified in sub-paragraph (h);

VINVCOM = Increase or decrease in the cost of inventory for telecommunications works during the period under consideration due to changes in the rates for relevant components as specified in sub-paragraph (h);

VINTGTESTCOM = Increase or decrease in the cost of integrated testing and commissioning of telecommunication worksof the Railway Project during the period under consideration dueto changes in the rates for relevant components as specified in sub- paragraph (h);

PCEQP, PELEX, PIC, PLB, POFC, and POTH are the percentages of communication equipment, electronics, PVC insulated cables, labor, optical fiber cables, and other materials respectively;

CEQPo = The wholesale price index as published by the Ministry of Commerce and Industry, Government of India (hereinafter called "WPI") for communication equipment;

CEQPi = The WPI for communication equipment for the month three months prior to the month to which the IPC relates;

ELEXo = The WPI for electronics for the month of the Base Date;

ELEXi = The WPI for electronics for the month three months prior to the month to which the IPC relates;

ICo = The WPI for PVC insulated cables for the month of the Base Date;

ICi = The WPI for PVC insulated cables for the month three months prior to the month to which the IPC relates;

LBo = The consumer price index for industrial workers for the [circle **** in the State of Karnataka], published by Labor Bureau, Ministry of Labor, Government of India, (hereinafter called "CPI") for the month of the Base Date;

LBi = The CPI for the month three months prior to the month to which the IPC relates;

OFCo = The WPI for fiber cables for the month of the Base Date;

OFCi = The WPI for fiber cables for the month three months priorto the month to which the IPC relates;

OTHo = The WPI for all commodities for the month of the Base Date; and

OTHi = The WPI for all commodities for the month three months prior to the month to which the IPC relates.

(a) The following percentages shall govern the price adjustment of theContract Price for signal-ling and telecommunication works:

	Signalling			Telecommunication		
Component	**Signallingworks**	**Signallinginventory**	**Integratedtestingandcommissioning**	**Telecommunicationworks**	**Telecominventory**	**Integratedtestingandcommissioning**
Electronics (PELEX)	***%	***%	-	***%	***%	-
(PELEX)						
Communication Equipment (PCEQP)	-	-	-	***%	***%	-
Optical Fibre Cable (POFC)	***%	-	-	***%	-	-
PVC Insulated Cable (PIC)	***%	-	-	***%	-	-
Labour (PLB)	***%	-	***%	***%	***%	***%
Other materials	***%	***%	***%	***%	***%	***%
Total	100%	100%	100%	100%	100%	100%

(a) The following expressions and meanings are assigned to the value of the work done for electrification works:

OHE = Value of work done for the completion of a stage under the item Overhead Equipment Work;

SP = Value of work done for the completion of a stage under the item Switching Posts;

TRANSBOO = Value of work done for the completion of a stage under the item Booster Transformer;

TRANSAUX = Value of work done for the completion of a stage under the item Auxiliary Transformer;

TSS = Value of work done for the completion of a stage under the item Traction Sub Station;

TLOH = Value of work done for the completion of a stage under the item High Voltage Transmission Line Overhead including monopole;

TLUG = Value of work done for the completion of a stage under the item Underground High Tension Cable Transmission Line;

BAY = Value of work done for the completion of a stage under the item Bay Augmentation work at Grid Sub-Station/Terminal arrangement at TSS;

SCADA = Value of work done for the completion of a stage underthe item SCADA;

ELEGWK = Value of work done for the completion of a stage under the item various electrical general services works;

MODHTPWRLINE = Value of work done for the completion of a stage under the item modification of HT power lines and crossings (raising of height);

MODHTLTOUG = Value of work done for the completion of a stage under the item modification of HT power lines and crossingsto underground (replacement by underground cabling);

MODLTLTOUG = Value of work done for the completion of a stage under the item modification of LT power lines and crossings to underground (replacement by underground cabling);

EXTNLTPWRSPLY = Value of work done for the completion of a stage under the item extension/augmentation of power supply forCLS work;

EXTNPWRSUPLY = Value of work done for the completion of a stage under the item extension/augmentation of general power supply;

MODELETRICAL = Value of work done for the completion of a stage under the item modification to existing electrical works;

INVELECTRICAL = Value of work done for the completion of a stage under the item inventory electrical;

SIGMOD = Value of work done for the completion of a stageunder the item Signaling System Modification;

INVSIG = Value of work done for the completion of a stage underthe item signaling inventory;

TESTSIG = Value of work done for the completion of a stage

under the item integrated testing and commissioning;

COMMOD = Value of work done for the completion of a stage under the item Telecommunications modifications;

INVCOM = Value of work done for the completion of a stage under the item telecommunication inventory;

TESTCOM = Value of work done for the completion of a stage under the item integrated testing and commissioning; and

CIVENG = Value of work done for the completion of a stage underthe item Civil Engineering works.

(a) Price adjustment for changes in cost for electrification works shall be paid in accordance with the following formula:

I. VOHE = 0.85 OHE x [PLB x (LBi – LBo)/LBo + PC x (Ci–Co)/Co + PSST x (SSTi – SSTo)/SSTo + PCU x (CUi – CUo)/CUo + PINS x (INSi – INSo)/ INSo];

II. VSP = 0.85 SP x [PLB x (LBi – LBo)/LBo + PC x (Ci – Co)/Co + PSWGR x (SWGRi – SWGRo)/SWGRo];

III. VTRANSBOO = 0.85 TRANSBOO x [PLB x (LBi –LBo)/LBo + PSST x (SSTi – SSTo)/SSTo + PTR x (TRi –TRo)/TRo];

IV. VTRANSAUX = 0.85 TRANSAUX x [PLB x (LBi –LBo)/LBo + PSST x (SSTi – SSTo)/SSTo + PTR x (TRi –TRo)/TRo];

V. VTSS = 0.85 TSS x [PLB x (LBi – LBo)/LBo + PTR (TRi – TRo)/TRo + PC x (Ci – Co)/ Co + PSST x (SSTi – SSTo)/SSTo + PSWGR x (SWGRi – SWGRo)/SWGRo];

VI. VTLOH = 0.85 TLOH x [PLB x (LBi – LBo)/LBo + PSSTx (SSTi – SSTo)/SSTo + PCOND x (CONDi – CONDo)/CONDo + PC x (Ci – Co)/ Co + PINS x (INSi – INSo)/ INSo + POTH x (OTHi – OTHo)OTHo];

VII. VTLUG = 0.85 TLUG x [PLB x (LBi – LBo)/LBo + PIC x(ICi

– ICo)/ICo];

VIII. VBAY = 0.85 BAY x [PLB x (LBi – LBo)/LBo + PSST x (SSTi – SSTo)/SSTo + PC x (Ci – Co)/ Co + PCU x (CUi – CUo)/CUo];

IX. VSCADA = 0.85 SCADA x [PLB x (LBi – LBo)/LBo + PELEX x (ELEXi – ELEXo)/ELEXo];

X. VELEGWK = 0.85 ELEGW x [PLB x (LBi – LBo)/LBo + POTH x (OTHi – OTHo)OTHo];

XI. VMODHTPWRLINE = 0.85 MODHTPWRLINE x [PLB x (LBi – LBo)/LBo + PSST x (SSTi – SSTo)/SSTo + POTH x (OTHi – OTHo)/OTHo];

XII. VMODHTLTOUG = x0.85 MODHTLTOUG x [PLB x (LBi – LBo)/LBo + PIC x (ICi – ICo)/ICo + POTH x (OTHi – OTHo)/OTHo];

XIII. VMODLTLTOUG = 0.85 MODLTLTOUG x [PLB x (LBi – LBo)/LBo + PIC x (ICi – ICo)/ICo + POTH x (OTHi – OTHo)/OTHo];

XIV. VEXTNLTPWRSPLY = 0.85 EXTNLTPWRSPLY x[PLB x (LBi – LBo)/LBo + POTH x (OTHi – OTHo)/OTHo];

XV. VEXTNPWRSUPLY = 0.85 EXTNPWRSUPLY x [PLB x (LBi – LBo)/LBo + POTH x (OTHi – OTHo) OTHo];

I. VMODELETRICAL = 0.85 MODELETRICAL x [PLB x (LBi – LBo)/LBo + POTH x (OTHi – OTHo)/OTHo];

II. INVELECTRICAL = 0.85 INVELECTRICAL x [POTH x (OTHi – OTHo)/OTHo];

III. VSIGMOD = 0.85 SIGMOD x [PLB x (LBi – LBo)/LBo + PELEX x (ELEXi – ELEXo)/ELEXo + PIC x (ICi – ICo)/ICo +POTH x (OTHi– OTHo)/OTHo];

IV. VINVSIG = 0.85 INVSIG x [POTH x (OTHi – OTHo)/OTHo];

V. VTESTSIG = 0.85 TESTSIG x [PLB x (LBi – LBo)/LBo + POTH x (OTHi – OTHo)/OTHo];

VI. VCOMMOD = 0.85 COMMOD x [PLB x (LBi – LBo)/LBo + PELEX x (ELEXi – ELEXo)/ELEXo + POFC x (OFCi – OFCo)/OFCo];

VII. VINVCOM = 0.85 INVCOM x [POTH x (OTHi – OTHo)/OTHo];

VIII. VTESTCOM = 0.85 TESTCOM x [PLB x (LBi –LBo)/LBo + POTH x (OTHi – OTHo)/OTHo]; and

IX. VCIVENG = 0.85 x VCIVENG x [PLB x (LBi – LBo)/LBo + PS x (Si – So)/So + PC x (Ci – Co)/ Co + POTH x(OTHi – OTHo)/ OTHo].

Where

VOHE = Increase or decrease in the cost of Over Head Equipmentand other related works during the period under consideration due to changes in the rates for relevant components as specified in sub-paragraph (k);

VSP = Increase or decrease in the cost of Switch Post and other related works during the period under consideration due to changes in the rates for relevant components as specified in sub- paragraph (k);

VTRANSBOO = Increase or decrease in the cost of booster transformer and other related works during the period under consideration due to changes in the rates for relevant componentsas specified in sub-paragraph (k);

VTRANSAUX = Increase or decrease in the cost of auxiliary transformer and other related works during the period under consideration due to changes in the rates for relevant componentsas specified in sub-paragraph (k);

VTSS = Increase or decrease in the cost of Traction Sub-Station and other related works during the period under consideration due to changes in the rates for relevant components as specified in sub-paragraph (k);

VTLOH = Increase or decrease in the cost of overhead transmission line and related works during the period under consideration due to changes in the rates for relevant components as specified in sub-paragraph (k);

VTLUG = Increase or decrease in the cost of underground high voltage transmission line and related works during the period under Consideration due to changes in the rates for relevantcomponents as specified in sub- paragraph (k);

VBAY = Increase or decrease in the cost of bay augmentation work at grid sub-station/ terminal arrangement at TSS and relatedworks during the period under consideration due to changes in therates for relevant components as specified in sub-paragraph (k);

VSCADA = Increase or decrease in the cost of SCADA and related works during the period under consideration due to changes in therates for relevant components as specified in sub-paragraph (k);

VELEGWK = Increase or decrease in the cost of various electrical general services works and related works during the period under consideration due to changes in the rates for relevant componentsas specified in

sub-paragraph (k);

VMODHTPWRLINE = Increase or decrease in the cost of modification of HT power lines and crossings (raising of height) and related works during the period under consideration due to changes in the rates for relevant components as specified in sub- paragraph (k);

VMODHTLTOUG = Increase or decrease in the cost of modification of HT power lines and crossings to underground (replacement by underground cabling) and related works during the period under consideration due to changes in the rates for relevant components as specified in sub-paragraph (k);

VMODLTLTOUG = Increase or decrease in the cost of modification of LT power lines and crossings to underground (replacement by underground cabling) and related works during the period under consideration due to changes in the rates for relevant components as specified in sub-paragraph (k);

VEXTNLTPWRSPLY = Increase or decrease in the cost of extension/augmentation of power supply for CLS work and related works during the period under consideration due to changes in therates for relevant components as specified in sub-paragraph (k);

VEXTNPWRSUPLY = Increase or decrease in the cost of extension/augmentation of general power supply and related works during the period under consideration due to changes in therates for relevant components as specified in sub-paragraph (k);

VMODELETRICAL = Increase or decrease in the cost of modification to existing electrical works and related works duringthe period under consideration due to changes in the rates for relevant components as specified in sub-paragraph (k);

VINVELECTRICAL = Increase or decrease in the cost of inventory electrical during the period under consideration due to changes in the rates for relevant components as specified in sub- paragraph (k);

VSIGMOD = Increase or decrease in the cost of signaling system modification and related works during the period under consideration due to changes in the rates for relevant componentsas specified in sub-paragraph (k);

VINVSIG = Increase or decrease in the cost of signaling inventory during the period under consideration due to changes inthe rates for relevant components as specified in sub-paragraph (k);

VTESTSIG = Increase or decrease in the cost of SCADE and related works during the period under consideration due to changes in the rates for relevant components as specified in

sub-paragraph (k);

VCOMMOD = Increase or decrease in the cost of communication and related works during the period under consideration due tochanges in the rates for relevant components as specified in sub- paragraph (k);

VINVCOM = Increase or decrease in the cost oftelecommunication inventory during the period under consideration due to changes in the rates for relevant componentsas specified in sub-paragraph (k);

VTESTCOM = Increase or decrease in the cost of integrated testing and commissioning and related works during the period under consideration due to changes in the rates for relevant components as specified in sub-paragraph (k);

VCIVENG = Increase or decrease in the cost of civil engineering and related works during the period under consideration due tochanges in the rates for relevant components as specified in sub- paragraph (k);

PC, PCOND, PCU, PELEX, PINS, PLB, POFC, PSWGR, PPC,

and PSST are the percentages of cement, conductor, copper wire, electronic items, insulators, labor, fiber optic cables, electrical switch gears, PVC insulated cable and structural steel respectively for the relevant item as specified in sub-paragraph (k);

Co = The wholesale price index as published by the Ministry of Commerce & Industry, Government of India (hereinafter called "WPI") for grey cement for the month of the Base Date;

Ci = The WPI for grey cement for the month three months prior to the month to which the IPC relates;

CONDo = the WPI conductors for the month of the Base Date;

CONDi = The WPI for conductors for the month three months prior to the month to which the IPC relates;

CUo = The WPI for copper (all types) for the month of the Base Date;

CUi = The WPI for copper (all types) for the month three months prior to the month to which the IPC relates;

ELEXo = The WPI for electronic items for the month of the Base Date;

ELEXi = The WPI for electronic items for the month threemonths prior to the month to which the IPC relates;

INSo = The WPI for insulators for the month of the Base Date;

INSi = The WPI for insulators for the month three months priorto the month to which the IPC relates;

LBo = The consumer price index for industrial workers for the [circle **** in the State of ***] published by Labor Bureau, Ministry of Labor, Government of India, (hereinafter called "CPI")for the month of the Base Date;

LBi = The CPI for the month three months prior to the month to which the IPC relates;

OFCo = The WPI for optical fiber cables for the month of the Base Date;

OFCi = The WPI for optical fiber cables for the month three

months prior to the month to which the IPC relates;

Oo = The WPI for all commodities for the month of the Base Date;

Oi = The WPI for all commodities for the month three months prior to the month to which the IPC relates;

PCo = The WPI for PVC insulated cable for the month of the Base Date;

PCi = The WPI for PVC insulated cable for the month three months prior to the month to which the IPC relates;

So = The WPI for steel (rods) for the month of theBase Date;

Si = The WPI for steel (rods) for the month three months prior tothe month to which the IPC relates;

SSTo = The WPI for structural steel for the month of the Base Date;

SSTi = The WPI for structural steel for the month three months prior to the month to which the IPC relates;

SWGRo = The WPI for electric switch gears for the month of the Base Date;

SWGRi = The WPI for electric switch gears for the month three months prior to the month to which the IPC relates;

TRo = The WPI for transformers for the month of the Base Date; and

TRi = The WPI for transformers for the month three months prior to the month to which the IPC relates.

(a)The following percentages shall govern the price adjustment of the Contract Price for

electrification works:

(i) For OHE, TSS, SP, Booster Transformer stations, Auxiliary-transformer stations:

Component	Over Head Equipment		Switch Posts except commissioning and charging	Booster Transformer Station	Tracking sub stations		Auxiliary transformer stations	OHE other works, commissioning and charging of TSS,SP, Booster Transformer Stations, Auxilliary
	Foundation, mast erection, bracket erection, insulators	Stringing of catenary and contact wire			Transformers	All works except transformers		
Labour (PLB)	***%		***%	***%		***%	***%	100%
Cement (PC)	***%		***%	-		***%	-	-
Structural steel(PSS	***%		-	***%		***%	***%	-
Insulators(PINS)	***%		-	-	-			
Copper wire(PCU)	-		-	-			***%	-
Transformer(PTR)	-		-	***%	100%	***%	-	-
Electrical Switch Gear(PSWGR)	-		***%	-				
Total	100%		100%	100%				

(I) For overhead transmission lines, underground high-tension cable transmission line, bay augmentation work at Grid Sub- station etc., various electrical general services works and modification of HT power lines and crossings (raising of height):

Component	Transmission lines overhead including monopole except commissioning	Underground high tension cable transmission line except commissioning	Bay augmentation work at grid sub-station/terminal arrangement at TSS	Various Modification of electrical lines and crossings(raising of height)	Modification of HT power lines and crossings (raising of height)	Commissioning of transmission lines overhead,underground high tension cable transmission line,bay augmentation work
Labour (PLB)	***%	***%	***%	***%	***%	100%
Structural steel	***%	-	***%	-	***%	-
Cement (PC)	***%	-	***%	-	-	-
Conductor (PCOND)	***%	-	-			
PVC Insulated Cable (PIC)	-	***%	-	-	-	-
Copper wire (PCU)	-	-	***%	-	-	-
Insulators (PINS)	***%	-	-	-	-	-
Other items (POTH)	***%	-	-	***%	***%	-
Total	100%	100%	100%	100%	100%	100%

(I) For SCADA, modification of HT power lines and crossings to underground (replacement by underground cabling), modificationof LT power lines and crossings to underground (replacement by underground cabling except commissioning, Extension/augmentation of power supply for CLS work, extension/ augmentation of general power supply, modification to existing electrical works:

Component	SCADA except com-missioning for the Division	Modification of HT power lines and crossings to under ground(replacement by underground cabling)except commissioning	Modification of LT power lines and crossings to underground(replacement by underground cabling except commissioning	Extension/augmentation of power supply for CLS work except commissioning	Extension/augmentation of general power supply	Modification to existing electrical works	Commissioning of SCADA,Modification of HT power lines,modification of LT power lines,and Extension/augmentation of power supply for CLS work
Labour (PLB)	***%	***%	***%	***%	***%	***%	100%
Electronics(PELEX)	**%	-	-	-	-	-	-
PVC Insulated CABLE(PIC)	-	***%	***%	*	*	*	-
Fibre Cable(POFC)	-	-	-	-	-	-	-
All other commodities (POTH)	***%	***%	***%	***%	***%	***%	-
Total	100%	100%	100%	100%	100%	100%	100%

(i) For modification of signaling works, modification of telecommunications works, inventory for electrification. Signaling and telecommunication works; and integrated testing andcommissioning of the electrification, signaling and telecommunication works

Component	Modification of Signalling Works	Modification of Telecommunication Works	Inventory for electrification signalling and telecommunication works.	Integrated testing and commissioning of electrification.signalling and telecommunication works.
Labour(PLB)	***%	***%	-	***%
Electronics(PELEX)	***%	***%	-	-
PVC Insulated Cable(PIC)	***%	***%	-	-
Fibre Cable (POFC)	-	***%	-	-
All other commodities (POTH	***%	***%	***%	***%
Total	100%	100%	100%	100%

(v) For Civil Engineering Works:

	Component			
	Labour (PLB)		***%	
	Steel (PS)		***%	
	Cement (PC)		***%	
	All other commodities(POTH)		***%	

(i) For Civil Engineering Works:

14.16.5 14.16.5

In case an IPC relates to a month which is within 3 (three) months from the Base Date, price adjustment is applicable on the basis of overhead cost and rate finalization.

Termination/Takeover by Engineer

15.1 Notice to Correct

If the Contractor fails to carry out any obligation under theContract, the Engineer may by notice require the Contractor to make good the failure and to remedy it within a specified reasonable time.

15.2 Termination/Takeover by Engineer

Termination is the extreme, In a democratic manner if performance of Contractor is not good reason whatever required to perform the activity of takeover and complete the job and sold out the Firm to save the Livelihood of Employee and utilization of Plant & Machinery.

In extreme cases after committee report termination activity should be taken.

The Engineer shall be entitled to terminate or Takeover the Contract if the Contractor:

a) fails to comply with Sub-Clause 4.2 [Performance Security] or with a notice under Sub-Clause 15.1 [Notice to Correct],

b) abandons the Works or otherwise plainly demonstrates the intention not to continue performance of his obligations underthe Contract,

c) without reasonable excuse fails: (I) to proceed with the Worksin accordance with Clause 8 [Commencement, Delays and Suspension], or (ii) to comply with a notice issued under Sub-Clause 7.5 [Rejection] or Sub Clause 7.6 [Remedial Work], within 28 days after receiving it,

d) subcontracts the whole of the Works or assigns the Contract without the required agreement,

e) becomes bankrupt or insolvent, goes into liquidation, has a receivingor administrationorder made against him, compounds with his creditors, or carries on business under a receiver, trustee or manager for the benefit of his creditors, orif any act is done or event occurs which (under applicable Laws)has a similar effect to any of these acts or events, or (f) gives or offers to give (directly or indirectly) to any person any bribe, gift, gratuity, commission or other thing of value, as an inducement or reward:

I. for doing or forbearing to do any action in relation to the Contract, or

II. for showing or forbearing to show favor or disfavor to anyperson in relation to the Contract, or if any of the Contractor'sPersonnel, agents or Subcontractors gives or offers to give (directly or indirectly) to any person any such inducement orreward as is described in this sub-paragraph

(f). However, lawful inducements and rewards to Contractor's Personnel shall not entitle termination.

In any of these events or circumstances, the Engineer may, upon giving 14 days' notice to the Contractor, takeover/terminate the Contract and expel the Contractor from the Site.

However, in the case of sub-paragraph (e) or (f), the Employer may by notice takeover/terminate the Contract immediately.

The Engineer's election to takeover/terminate the Contract shall not prejudice any other rights of the Employer, under the Contract or otherwise.

The Contractor shall then leave the Site and deliver any required Goods, all Contractor's Documents, and other design documents made by or for him, to the Engineer.

However, the Contractor shall use his best efforts to comply immediately with any reasonable instructions included in thenotice

i. for the assignment of any subcontract, and
ii. for the protection of life or property or for the safety of the Works.
    After takeover/termination, the Engineer may complete the Works and/or arrange for any other entities to do so.

The Engineer and these entities may then use any Goods, Contractor's Documents and other design documents made by oron behalf of the Contractor.

The Engineer shall then give notice that the Contractor's Equipment and Temporary Works will be released to the Contractor at or near the Site.

The Contractor shall promptly arrange their removal, at the risk and cost of the Contractor.

However, if by this time the Contractor has failed to make a payment due to the Engineer, these items may be sold by the Engineer in order to recover this payment.

Any balance of the proceeds shall then be paid to the Contractor.

15.3 Valuation at Date of Termination

As soon as practicable after a notice of termination under Sub-Clause 15.2 [Termination by Employer] has taken effect, the Engineer shall proceed in accordance with Sub-Clause 3.5 [Determinations] to agree or determine the value of the Works, Goods and Contractor's Documents, and any other sums due to theContractor for work executed in accordance with the Contract.

15.4 Payment after Termination

After a notice of termination under Sub-Clause 15.2 [Termination by Employer] has taken effect, the Engineer may:

a. proceed in accordance with Sub-Clause 2.5 [Employer's Claims],
b. withhold further payments to the Contractor until the costs of execution, completion and remedying of any defects, damages for delay in completion (if any), and all other costs incurred by the Engineer, have been established, and/or
c. recover from the Contractor any losses and damages incurredby the Employer and any extra costs of completing the Works,after allowing for any sum due to the Contractor under Sub- Clause 15.3 [Valuation at Date of Termination].

After recovering any such losses, damages and extra costs, the Engineer shall pay any balance to the Contractor.

15.5 Engineer's Entitlement to Termination

The Engineer shall be entitled to terminate the Contract, at any time for the Engineer's convenience, by giving notice of such termination to the Contractor. The termination shall take effect 28 days after the later of the dateson which the Contractor receives this notice or the Engineer returns the Performance Security.

The Engineer shall not terminate the Contract under this Sub- Clause in order to execute the Works himself or to arrange for the Works to be executed by another contractor. After this termination,the Contractor shall proceed in accordance with Sub Clause 16.3 [Cessation of Work and Removal of Contractor's Equipment] and shall be paid in accordance with Sub-Clause 19.6 [Optional Termination, Payment and Release].

Suspension and Termination by Contractor

16.1 Contractor's Entitlement to Suspend Work

If the Engineer fails to certify in accordance with Sub-Clause 14.6 [Issue of Interim Payment Certificates] or the Employer fails to comply with Sub-Clause 2.4 [Employer's Financial Arrangements] or Sub-Clause 14.7 [Payment], the Contractor may, after giving not less than 21 days' notice to the Employer, suspend work (or reduce the rate of work) unless and until the Contractor has received the Payment Certificate, reasonable evidence or payment, as the case may be and as described in the notice.

The Contractor's action shall not prejudice his entitlements to financing charges under Sub-Clause 14.8 [Delayed Payment] andto termination under Sub-Clause 16.2 [Termination by Contractor].

If the Contractor subsequently receives such Payment Certificate, evidence or payment (as described in the relevant Sub-Clause andin the above notice) before giving a notice of termination, the Contractor shall resume normal working as soon as is reasonablypracticable.

If the Contractor suffers delay and/or incurs Cost as a result of suspending work (or reducing the rate of work) in accordance with this Sub-Clause, the Contractor shall give notice to the

Engineer and shall be entitled subject to Sub-Clause 20.1 [Contractor's Claims] to: an extension of time for any such delay, if completion is or will be delayed, under Sub-Clause 8.4 [Extension of Time for Completion], and

a) payment of any such Cost-plus reasonable profit, which shall be included in the Contract Price.

b) After receiving this notice, the Engineer shall proceed in accordance with Sub Clause 3.5 [Determinations] to agree or determine these matters.

16.2 Termination by Contractor

The Contractor shall be entitled to terminate the Contract if:

a. the Contractor does not receive the reasonable evidence within 42 days after giving notice under Sub-Clause 16.1 [Contractor's Entitlement to Suspend Work] in respect of a failure to comply with Sub-Clause 2.4 [Employer's Financial Arrangements],

b. the Engineer fails, within 56 days after receiving a Statement and supporting documents, to issue the relevant Payment certificate,

c. the Contractor does not receive the amount due under an Interim Payment Certificate within 42 days after the expiry ofthe time stated in Sub-Clause 14.7 [Payment] within which payment is to be made (except for deductions in accordance with Sub-Clause 2.5 [Employer's Claims]),

d. the Employer substantially fails to perform his obligations under the Contract,

e. the Employer fails to comply with Sub-Clause 1.6 [Contract Agreement] or Sub-Clause 1.7 [Assignment],

f. a prolonged suspension affects the whole of the Works as described in Subclause 8.11 [Prolonged Suspension], or

g. the Employer becomes bankrupt or insolvent, goes into liquidation, has a receiving or administration order madeagainst him, compounds with his creditors, or carries onbusiness under a receiver, trustee or manager for the benefit ofhis creditors, or if any act is done or event occurs which (under applicable Laws) has a similar effect to any of these acts or events.

In any of these events or circumstances, the Contractor may, upon giving 14 days' notice to the Employer, terminate the Contract. However, in the case of subparagraph (f) or (g), the Contractor may by notice terminate the Contract immediately.

The Contractor's election to terminate the Contract shall not prejudice any other rights of the Contractor, under the Contract or otherwise.

16.3 Cessation of Work and Removal of Contractor's Equipment

After a notice of termination under Sub-Clause 15.5 [Employer's Entitlement to Termination], Sub-Clause 16.2 [Termination by Contractor] or Sub-Clause 19.6 [Optional Termination, Payment and Release] has taken effect, the Contractor shall promptly:

a) cease all further work, except for such work as may have been instructed by the Engineer for the protection of life or property or for the safety of the Works,

b) hand over Contractor's Documents, Plant, Materials and other work, for which the Contractor has received payment, and (c) remove all other Goods from the Site, except as necessary for safety, and leave the Site.

16.4 Payment on Termination

After a notice of termination under Sub-Clause 16.2 [Terminationby Contractor] has taken effect, the Employer shall promptly:

a. return the Performance Security to the Contractor,

b. pay the Contractor in accordance with Sub-Clause 19.6 [Optional Termination, Payment and Release], and

c. pay to the Contractor the amount of any loss of profit or otherloss or damage sustained by the Contractor as a result of this termination.

Risk and Responsibility

17.1 Indemnities

Meaning is security or protection against a loss or other financial burden.

The Contractor shall indemnify and hold harmless the Engineer,the Engineer's Personnel, and their respective agents, against and from all claims, damages, losses and expenses (including legal fees and expenses) in respect of:

a. bodily injury, sickness, disease or death, of any person whatsoever arising out of or in the course of or by reason of the Contractor's design (if any), the execution and completionof the Works and the remedying of any defects, unless attributable to any negligence, willful act or breach of the Contract by the Employer, the Employer's Personnel, or any of their respective agents, and

b. damage to or loss of any property, real or personal (other thanthe Works), to the extent that such damage or loss:

i. arises out of or in the course of or by reason of the Contractor's design (if any), the execution and completion ofthe Works and the remedying of any defects, and

ii. is attributable to any negligence, willful act or breach of the Contract by the Contractor, the Contractor's Personnel, their respective agents, or anyone directly or indirectly employed by any of them.

The Engineer shall indemnify and hold harmless the Contractor, the Contractor's Personnel, and their respective agents, against and from all claims, damages, losses and expenses (including legal fees and expenses) in respect of

1. bodily injury, sickness, disease or death, which is attributable to any negligence, willful act or breach of the Contract by the Employer, the Employer's Personnel, or any of their respective agents, and
2. the matters for which liability may be excluded from insurance cover, as described in sub-paragraphs (d)(I), (ii) and (iii) of Sub-Clause 18.3 [Insurance Against Injury to Persons and Damageto Property].

17.2 Contractor's Care of the Works

The Contractor shall take full responsibility for the care of the Works and Goods from the Commencement Date until theTaking-Over Certificate is issued (or is deemed to be issued under Sub-Clause 10.1 [Taking Over of the Works and

Sections]) for the Works, when responsibility for the care of the Works shall pass to the Engineer, but in maintenance period responsibility should continue with the Contractor's.

If a Taking-Over Certificate is issued (or is so deemed to be issued) for any Section or part of the Works, responsibility for the care of the Section or part shall then pass to the Engineer.

After responsibility has accordingly passed to the Engineer, the Contractor shall take responsibility for the care of any work whichis outstanding on the date stated in a Taking-Over Certificate, until this outstanding work has been completed.

If any loss or damage happens to the Works, Goods or Contractor's Documents during the period when the Contractor isresponsible for their care, from any cause not listed in Sub-Clause

17.3 [Employer's Risks], the Contractor shall rectify the loss or damage at the Contractor's risk and cost, so that the Works, Goods and

Contractor's Documents conform with the Contract.

The Contractor shall be liable for any loss or damage caused by any actions performed by the Contractor after a Taking-Over Certificate has been issued.

The Contractor shall also be liable for any loss or damage which occurs after a Taking Over Certificate has been issued and which arose from a previous event for which the Contractor was liable.

17.3 Engineer's Risks

The risks referred to in Sub-Clause 17.4 below are:

a) war, hostilities (whether war be declared or not), invasion, actof foreign enemies,
b) rebellion, terrorism, revolution, insurrection, military or usurped power, or civil war, within the Country,
c) riot, commotion or disorder within the Country by persons other than the Contractor's Personnel and other employees of the Contractor and Subcontractors,
d) munitions of war, explosive materials, ionizing radiation or contamination by radio-activity, within the Country, except as may be attributable to the Contractor's use of such munitions, explosives, radiation or radio-activity,
e) pressure waves caused by aircraft or other aerial devices travelling at sonic or supersonic speeds,
f) use or occupation by the Engineer of any part of the Permanent Works, except as may be specified in the Contract,
g) design of any part of the Works by the Engineer's Personnel or by others for whom the Engineer is responsible, and
h) any operation of the forces of nature which is Unforeseeable or

against which an experienced contractor could not reasonably have been expected to have taken adequate preventative precautions.

17.4 Consequences of Engineer's Risks

If and to the extent that any of the risks listed in Sub-Clause 17.3 above results in loss or damage to the Works, Goods or Contractor's Documents, the Contractor shall promptly give notice to the Engineer and shall rectify this loss or damage to the extent required by the Engineer.

If the Contractor suffers delay and/or incurs Cost from rectifyingthis loss or damage, the Contractor shall give a further notice to the Engineer and shall be entitled subject to Sub-Clause 20.1 [Contractor's Claims] to:

a. an extension of time for any such delay, if completion is or will be delayed, under Sub-Clause 8.4 [Extension of Time for Completion], and

b. payment of any such Cost, which shall be included in the Contract Price. In the case of sub-paragraphs (f) and (g) of Sub-Clause 17.3 [Employer's Risks], reasonable profit on the Cost shall also be included.

After receiving this further notice, the Engineer shall proceed in accordance with Sub Clause 3.5 [Determinations] to agree or determine these matters.

17.5 Intellectual and Industrial Property Rights

In this Sub-Clause, "infringement" means an infringement (or alleged infringement) of any patent, registered design, copyright,trade mark, trade name, trade secret or other intellectual or industrial property right relating to the Works; and "claim" meansa claim (or proceedings pursuing a claim) alleging an infringement. Whenever a

Party does not give notice to the other Party of any claim within 28 days of receiving the claim, the first Party shall bedeemed to have waived any right to indemnity under this Sub- Clause. The Engineer shall indemnify and hold the Contractor harmless against and from any claim alleging an infringement which is or was:

a) an unavoidable result of the Contractor's compliance with the Contract, or

b) a result of any Works being used by the Engineer:

i. for a purpose other than that indicated by, or reasonably to be inferred from, the Contract, or

ii. in conjunction with anything not supplied by the Contractor,unless such use was disclosed to the Contractor prior to the Base Date or is stated in the Contract.

The Contractor shall indemnify and hold the Engineer harmless against and from any other claim which arises out of or in relationto

I. the manufacture, use, sale or import of any Goods, or

ii any design for which the Contractor is responsible. If a Party is entitled to be indemnified under this Sub-Clause, the indemnifying Party may (at its cost) conduct negotiations forthe settlement of the claim, and any litigation or arbitration which may arise from it.

The other Party shall, at the request and cost of the indemnifying Party, assist in contesting the claim.

This other Party (and its Personnel) shall not make any admission which might be prejudicial to the indemnifying Party, unless the indemnifying Party failed to take over the conduct of anynegotiations, litigation or arbitration upon being requested to doso by such other Party.

17.6 Limitation of Liability

Neither Party shall be liable to the other Party for loss of use of any Works, loss of profit, loss of any contract or for any indirect or consequential loss or damage which may be suffered by the other Party in connection with the Contract, other than under

Sub-Clause 16.4 [Payment on Termination] and Sub-Clause 17.1 [Indemnities].

The total liability of the Contractor to the Engineer, under or in connection with the Contract other than under Sub-Clause 4.19 [Electricity, Water and Gas], Subclause 4.20 [Engineer's Equipment and Free-Issue Material], Sub-Clause 17.1

[Indemnities] and Sub-Clause 17.5 [Intellectual and Industrial Property Rights], shall not exceed the sum stated in the Particular Conditions or (if a sum is not so stated) the Accepted Contract Amount.

This Sub-Clause shall not limit liability in any case of fraud, deliberate default or reckless misconduct by the defaulting Party.

Insurance

18.1 General Requirements for Insurances

In this Clause, "insuring Party" means, for each type of insurance,the Party responsible for effecting and maintaining the insurance specified in the relevant Sub-Clause.

Wherever the Contractor is the insuring Party, each insurance shall be affected with insurers and in terms approved by the Engineer.

These terms shall be consistent with any terms agreed by both Parties before the date of the Letter of Acceptance.

This agreement of terms shall take precedence over the provisions of this Clause.

Wherever the Engineer is the insuring Party, each insurance shallbe affected with insurers and in terms consistent with the details annexed to the Particular Conditions.

If a policy is required to indemnify joint insured, the cover shall apply separately to each insured as though a separate policy had been issued for each of the joint insured.

If a policy indemnifies additional joint insured, namely in additionto the insured specified in this Clause,

I.) the Contractor shall act under the policy on behalf of these additional joint insured except that the Engineer shall act for Engineer's Personnel,

ii.) additional joint insured shall not be entitled to receive payments directly from the insurer or to have any other direct dealings with the insurer, and

iii.) the insuring Party shall require all additional joint insured to comply with the conditions stipulated in the policy.

Each policy insuring against loss or damage shall provide for payments to be made in the currencies required to rectify the lossor damage.

Payments received from insurers shall be used for the rectification of the loss or damage. The relevant insuring Party shall, within the respective periods stated in the Appendix to Tender (calculated from the Commencement Date), submit to the other Party:

a) evidence that the insurances described in this Clause have been affected, and

b) copies of the policies for the insurances described in Sub- Clause 18.2 [Insurance for Works and Contractor's Equipment] and Sub-Clause 18.3 [Insurance against Injury toPersons and Damage to Property].

When each premium is paid, the insuring Party shall submit evidence of payment to the other Party.

Whenever evidence or policies are submitted, the insuring Party shall also give notice to the Engineer.

Each Party shall comply with the conditions stipulated in each ofthe insurance policies.

The insuring Party shall keep the insurers informed of any relevant changes to the execution of the Works and ensure that insurance is maintained in accordance with this Clause.

Neither Party shall make any material alteration to the terms of any insurance without the prior approval of the other Party.

If an insurer makes (or attempts to make) any alteration, the Party first notified by the insurer shall promptly give notice to the otherParty.

If the insuring Party fails to effect and keep in force any of the insurances it is required to effect and maintain under the Contract,or fails to provide satisfactory evidence and copies of policies in

accordance with this Sub-Clause, the other Party may (at its option and without prejudice to any other right or remedy) effect insurance for the relevant coverage and pay the premiums due.

The insuring Party shall pay the amount of these premiums to the other Party, and the Contract Price shall be adjusted accordingly.

Nothing in this Clause limits the obligations, liabilities or responsibilities of the Contractor or the Engineer, under the other terms of the Contract or otherwise.

Any amounts not insured or not recovered from the insurers shall be borne by the Contractor and/or the Engineer in accordance with these obligations, liabilities or responsibilities. However, if the insuring Party fails to effect and keep in force an insurance which is available and which it is required to effect and maintain under the Contract, and the other Party neither approves the omission nor effects insurance for the coverage relevant to this default, any moneys which should have been recoverable under this insurance shall be paid by the insuring Party.

Payments by one Party to the other Party shall be subject to Sub-Clause 2.5 [Employer's Claims] or Sub-Clause 20.1 [Contractor's Claims], as applicable.

18.2 Insurance for Works and Contractor's Equipment

The insuring Party shall insure the Works, Plant, Materials and Contractor's Documents for not less than the full reinstatement cost including the costs of demolition, removal of debris and professional fees and profit.

This insurance shall be effective from the date by which the evidence is to be submitted under sub-paragraph

(a) of Sub-Clause 18.1 [General Requirements for Insurances], until the date of issue of the Taking-Over Certificate for the Works.

The insuring Party shall maintain this insurance to provide cover until the date of issue of the Performance Certificate, for loss or damage for which the Contractor is liable arising from a cause occurring prior to the issue of the Taking-Over Certificate, and forloss or damage caused by the Contractor in the course of any otheroperations (including those under Clause 11 [Defects Liability]).

The insuring Party shall insure the Contractor's Equipment for not less than the full replacement value, including delivery to Site.

For each item of Contractor's Equipment, the insurance shall be effective while it is being transported to the Site and until it is nolonger required as Contractor's Equipment. Unless otherwise stated in the Particular Conditions, insurances under this Sub- Clause:

a. shall be affected and maintained by the Contractor as insuring Party,
b. shall be in the joint names of the Parties, who shall be jointly entitled to receive payments from the insurers, payments being held or allocated between the Parties for the sole purpose of rectifying the loss or damage,
c. shall cover all loss and damage from any cause not listed in Sub-Clause 17.3 [Employer's Risks],
d. shall also cover loss or damage to a part of the Works which is attributable to the use or occupation by the Engineer of another part of the Works, and loss or damage from the risks listed in sub-paragraphs (c), (g) and (h) of Sub Clause 17.3 [Employer's Risks],

excluding (in each case) risks which are not insurable at commercially reasonable terms, with deductibles per occurrence of not more than the amount statedin the Appendix to Tender (if an amount is not so stated, thissub-paragraph (d) shall not apply), and

e. may however exclude loss of, damage to, and reinstatement of:

i. a part of the Works which is in a defective condition due to a defect in its design, materials or workmanship (but cover shall include any other parts which are lost or damaged as a direct result of this defective condition and not as described in sub- paragraph (ii) below),

ii. a part of the Works which is lost or damaged in order to reinstate any other part of the Works if this other part is in a defective condition due to a defect in its design, materials or workmanship,

iii. a part of the Works which has been taken over by the Engineer, except to the extent that the Contractor is liable forthe loss or damage, and

iv. Goods while they are not in the Country, subject to Sub- Clause 14.5 [Plant and Materials intended for the Works].

If, more than one year after the Base Date, the cover described insub-paragraph (d) above ceases to be available at commercially reasonable terms, the Contractor shall (as insuring Party) give notice to the Engineer, with supporting particulars.

The Engineer shall then

i. be entitled subject to Sub-Clause 2.5 [Employer's Claims] to payment of an amount equivalent to such commercially reasonable terms as the Contractor should have expected to have paid for such cover, and

ii. be deemed, unless he obtains the cover at commerciallyreasonable terms, to have approved the omission under Sub- Clause 18.1 [General Requirements for Insurances].

18.3 Insurance against Injury to Persons and Damage to Property:

The insuring Party shall insure against each Party's liability for any loss, damage, death or bodily injury which may occur to any physical property (except things insured under Sub-Clause 18.2 [Insurance for Works and Contractor's Equipment]) or to any person (except persons insured under Sub-Clause 18.4 [Insurancefor Contractor's Personnel]), which may arise out of the Contractor's performance of the Contract and occurring before theissue of the Performance Certificate.

This insurance shall be for a limit per occurrence of not less thanthe amount stated in the Appendix to Tender, with no limit on the number of occurrences.

If an amount is not stated in the Appendix to Tender, this Sub-Clause shall not apply.

Unless otherwise stated in the Particular Conditions, the insurances specified in this Sub-Clause:

a. shall be affected and maintained by the Contractor as insuring Party,
b. shall be in the joint names of the Parties,
c. shall be extended to cover liability for all loss and damage tothe Employer's property (except things insured under Sub- Clause 18.2) arising out of the Contractor's performance of the Contract, and
d. may however exclude liability to the extent that it arises from:

i. the Engineer's right to have the Permanent Works executed on, over, under, in or through any land, and to occupy this land for the Permanent Works,

ii. damage which is an unavoidable result of the Contractor's obligations to execute the Works and remedy any defects, and

iii. a cause listed in Sub-Clause 17.3 [Employer's Risks], except to the extent that cover is available at commercially reasonable terms.

18.4 Insurance for Contractor's Personnel

The Contractor shall affect and maintain insurance against liabilityfor claims, damages, losses and expenses (including legal fees and expenses) arising from injury, sickness, disease or death of any person employed by the Contractor or any other of the Contractor's Personnel.

The Employer and the Engineer shall also be indemnified under the policy of insurance, except that this insurance may excludelosses and claims to the extent that they arise from any act orneglect of the Employer or of the Employer's Personnel.

The insurance shall be maintained in full force and effect duringthe whole time that these personnel are assisting in the execution of the Works.

For a Subcontractor's employees, the insurance may be affected by the Subcontractor, but the Contractor shall be responsible for compliance with this Clause.

Force Majeure

19.1 Definition of Force Majeure

In this Clause, "Force Majeure" means an exceptional event or circumstance:

a. which is beyond a Party's control,

b. which such Party could not reasonably have provided against before entering into the Contract,

c. which, having arisen, such Party could not reasonably have avoided or overcome, and

d. which is not substantially attributable to the other Party. Force Majeure may include, but is not limited to, exceptional events or circumstances of the kind listed below, so long as conditions (a) to (d) above are satisfied:

i. war, hostilities (whether war be declared or not), invasion, actof foreign enemies,

ii. rebellion, terrorism, revolution, insurrection, military or usurped power, or civil war,

iii. riot, commotion, disorder, strike or lockout by persons other than the Contractor's Personnel and other employees of the Contractor and Subcontractors,

iv. munitions of war, explosive materials, ionizing radiation or contamination by radio-activity, except as may be attributable to the Contractor's use of such munitions, explosives, radiation or radio-activity, and

v. natural catastrophes such as earthquake, hurricane, typhoon or volcanic activity.

19.2 Notice of Force Majeure

If a Party is or will be prevented from performing any of its obligations under the Contract by Force Majeure, then it shall give notice to the other Party of the event or circumstances constituting the Force Majeure and shall specify the obligations, the performance of which is or will be prevented.

The notice shall be given within 14 days after the Party became aware, or should have become aware, of the relevant event or circumstance constituting Force Majeure.

The Party shall, having given notice, be excused performance of such obligations for so long as such Force Majeure prevents itfrom performing them.

Notwithstanding any other provision of this Clause, Force Majeure shall not apply to obligations of either Party to make payments tothe other Party under the Contract.

19.3 Duty to Minimize Delay

Each Party shall at all times use all reasonable endeavors to minimize any delay in the performance of the Contract as a resultof Force Majeure.

A Party shall give notice to the other Party when it ceases to be affected by the Force Majeure.

19.4 Consequences of Force Majeure

If the Contractor is prevented from performing any of hisobligations under the Contract by Force Majeure of which notice has been given under Sub-Clause 19.2 [Notice of Force Majeure], and suffers delay and/or incurs Cost by reason of such Force Majeure, the Contractor shall be entitled subject to Sub-Clause

20.1 [Contractor's Claims] to:

a. an extension of time for any such delay, if completion is or will be delayed, under Sub-Clause 8.4 [Extension of Time for Completion], and

b. if the event or circumstance is of the kind described in sub-paragraphs (I) to (iv) of Sub-Clause 19.1 [Definition of Force Majeure] and, in the case of subparagraphs (ii) to (iv), occursin the Country, payment of any such Cost. After receiving thisnotice, the Engineer shall proceed in accordance with Sub Clause 3.5 [Determinations] to agree or determine these matters.

19.5 19.5 Force Majeure Affecting Subcontractor

If any Subcontractor is entitled under any contract or agreement relating to the Works to relief from force majeure on termsadditional to or broader than those specified in this Clause, such additional or broader force majeure events or circumstances shallnot excuse the Contractor's non-performance or entitle him to relief under this Clause.

19.6 19.6 Optional Termination, Payment and Release

If the execution of substantially all the Works in progress is prevented for a continuous period of 84 days by reason of Force Majeure of which notice has been given under Sub-Clause 19.2 [Notice of Force Majeure], or for multiple periods which total more than 140 days due to the same notified Force Majeure, theneither Party may give to the other Party a notice of termination ofthe Contract. In this event, the termination shall take effect 7 daysafter the notice is given, and the Contractor shall proceed in accordance with Sub-Clause 16.3 [Cessation of Work and Removal of Contractor's Equipment].

Upon such termination, the Engineer shall determine the value ofthe work done and issue a Payment Certificate which shall include:

a. the amounts payable for any work carried out for which a price is stated in the Contract;

b. the Cost of Plant and Materials ordered for the Works which have been delivered to the Contractor, or of which the Contractor is liable to accept delivery: this Plant and Materialsshall become the property of (and be at the risk of) the Employer when paid for by the Employer, and the Contractor shall place the same at the Employer's disposal;

c. any other Cost or liability which in the circumstances was reasonably incurred by the Contractor in the expectation of completing the Works;

d. the Cost of removal of Temporary Works and Contractor's Equipment from the Site and the return of these items to the Contractor's works in his country (or to any other destinationat no greater cost); and

e. the Cost of repatriation (The act or process of restoring or returning someone or something to the country of origin, allegiance, or citizenship) of the Contractor's staff and labor employed wholly in connection with the Works at the date of termination.

19.7 Release from Performance under the Law

Notwithstanding any other provision of this Clause, if any event or circumstance outside the control of the Parties (including, but not limited to, Force Majeure) arises which makes it impossible or unlawful for either or both Parties to fulfil its or their contractual obligations or which, under the law governing the Contract, entitles the

Parties to be released from further performance of theContract, then upon notice by either Party to the other Party of such event or circumstance:

a. the Parties shall be discharged from further performance, without prejudice to the rights of either Party in respect of any previous breach of the Contract, and

b. the sum payable by the Employer to the Contractor shall be the same as would have been payable under Sub-Clause 19.6 [Optional Termination, Payment and Release] if the Contracthad been terminated under Sub-Clause 19.6.

Claim, Disputes and Arbitration

20.1 Contractor's Claims

If the Contractor considers himself to be entitled to any extensionof the Time for Completion and/or any additional payment, under any Clause of these Conditions or otherwise in connection with the Contract, the Contractor shall give notice to the Engineer, describing the event or circumstance giving rise to the claim.

The notice shall be given as soon as practicable, and not later than28 days after the Contractor became aware, or should have become aware, of the event or circumstance.

If the Contractor fails to give notice of a claim within such a period of 28 days, the Time for Completion shall not be extended, the Contractor shall not be entitled to additional payment, and the Engineer shall be discharged from all liability in connection withthe claim.

Otherwise, the following provisions of this Sub-Clause shall apply.

The Contractor shall also submit any other notices which are required by the Contract, and supporting particulars for the claim,all as relevant to such event or circumstance.

The Contractor shall keep such contemporary records as may be necessary to substantiate any claim, either on the Site or at another location acceptable to the Engineer.

Without admitting the Engineer's liability, the Engineer may, after receiving any notice under this Sub-Clause, monitor the record-keeping and/or instruct the Contractor to keep further contemporary records.

The Contractor shall permit the Engineer to inspect all these records and shall (if instructed) submit copies to the Engineer.

Within 42 days after the Contractor became aware (or should have become aware) of the event or circumstance giving rise to the claim, or within such other period as may be proposed by the Contractor and approved by the Engineer, the Contractor shall send to the Engineer a fully detailed claim which includes full supporting particulars of the basis of the claim and of the extension of time and/or additional payment claimed.

If the event or circumstance giving rise to the claim has a continuing effect:

a. this fully detailed claim shall be considered as interim.
b. the Contractor shall send further interim claims at monthly intervals, giving the accumulated delay and/or amount claimed, and such further particulars as the Engineer may reasonably require; and
c. the Contractor shall send a final claim within 28 days after the end of the effects resulting from the event or circumstance, or within such other period as may be proposed by the Contractor and approved by the Engineer.
d. Within 42 days after receiving a claim or any further particulars supporting a previous claim, or within such other period as may be proposed by the Engineer and approved by the Contractor, the Engineer shall respond with approval, or with disapproval and detailed comments. He may also request any necessary further particulars, but shall nevertheless give his response on the principles of the claim within such time.

Each Payment Certificate shall include such amounts for any claim as have been reasonably substantiated as due under the relevant provision of the Contract. Unless and until the particulars suppliedare sufficient to substantiate the whole of the claim, the Contractorshall only be entitled to payment for such part of the claim as he has been able to substantiate.

The Engineer shall proceed in accordance with Sub-Clause 3.5 [Determinations] to agree or determine (I) the extension (if any) ofthe Time for Completion (before or after its expiry) in accordancewith Sub-Clause 8.4 [Extension of Time for Completion], and/or

(ii) the additional payment (if any) to which the Contractor is entitled under the Contract.

The requirements of this Sub-Clause are in addition to those of any other Sub-Clause which may apply to a claim.

If the Contractor fails to comply with this or another Sub Clause in relation to any claim, any extension of time and/or additional payment shall take account of the extent (if any) to which the failure has prevented or prejudiced proper investigation of the claim, unless the claim is excluded under the second paragraph ofthis Sub-Clause.

20.2 Appointment of the Dispute Adjudication Board

Disputes shall be adjudicated by a DRC in accordance with Sub- Clause 20.4 [Obtaining Dispute Resolution Committee Decision].

The Parties shall jointly appoint a DRC by the date stated in the Appendix to Tender.

The DRC shall comprise, as stated in the Appendix to Tender,either three or six suitably qualified persons ("the members").

If the number is not so stated and the Parties do not agree otherwise, the DRC shall comprise six persons. If the DRC is to comprise six persons, each Party shall nominate two members for the approval of the other Party.

The Parties shall consult all these members and shall agree uponthe sixth member, who shall be appointed to act as chairman, compulsorily senior member of Engineer.

However, if a list of potential members is included in the Contract,the members shall be selected from those on the list, other than anyone who is unable or unwilling to accept appointment to the DRC.

The agreement between the Parties and either the sole member ("adjudicator") or each of the three members shall incorporate by reference the General Conditions of Dispute Adjudication Agreement contained in the Appendix to these General Conditions, with such amendments as are agreed between them.

The terms of the remuneration of either the sole member or each ofthe six members, including the remuneration of any expert whomthe DRC consults, shall be mutually agreed upon by the Parties when agreeing the terms of appointment.

Each Party shall be responsible for paying one-half of this remuneration.

If at any time the Parties so agree, they may jointly refer a matterto the DRC for it to give its opinion.

Neither Party shall consult the DRC on any matter without the agreement of the other Party.

If at any time the Parties so agree, they may appoint a suitably qualified person or persons to replace (or to be available to replace) any one or more members of the DRC. Final decision is with the ENGINEER.

Unless the Parties agree otherwise, the appointment will come into effect if a member declines to act or is unable to act as a result of death, disability, resignation or termination of appointment. If anyof these circumstances occurs and no such replacement is available, a replacement shall be appointed in the same manner as the replaced person was required to have been nominated or agreed upon, as described in this Sub-Clause.

The appointment of any member may be terminated by mutual agreement of three Parties, but not by the Engineer or the Contractor acting alone.

Unless otherwise agreed by three Parties, the appointment of the DRC (including each member) shall expire when the discharge referred to in Sub-Clause 14.12 [Discharge] shall have become effective.

20.3 Failure to Agree Dispute Adjudication Board

If any of the following conditions apply, namely:

a) the Parties fail to agree upon the appointment of the sole member of the DRC by the date stated in the first paragraph of Sub-Clause 20.2, [Appointment of the Dispute Resolution Committee]

b) either Party fails to nominate the member (for approval by theother Party) of a DRC of six persons by such date,

c) the Parties fail to agree upon the appointment of the sixth member (to act as chairman) of the DRC by such date, or

d) the Parties fail to agree upon the appointment of a replacement person within 42 days after the date on which the sole memberor one of the six members declines to act or is unable to act as a result of death, disability, resignation or termination of appointment, then the appointing entity or official named in the Appendix to Tender shall, upon the request of either or both of the Parties and after due consultation with both Parties,appoint this member of the DRC.

This appointment shall be final and conclusive. Each Party shall be responsible for paying one-half of the remuneration of the appointing entity or official.

20.4 Obtaining Dispute Resolution Committee's Decision

Engineer is responsible to deploy the DRC for 7 days to every project at every 90 days under leadership if Senior Engineer above

75 years. Report should submit site itself with complete presentation of clear view and all-party Engineer, Contractor and Authority has to follow the same.

In case suppose any dispute arises in between the 90 days, Engineer is responsible to deploy immediately the team of DRC to resolve the issue either on amicable settlement or committee will work like arbitrator to resolve the issue.

If a dispute (of any kind whatsoever) arises between the Parties in connection with, or arising out of, the Contract or the execution ofthe Works, including any dispute as to any certificate, determination, instruction, opinion or valuation of the Engineer, either Party may refer the dispute in writing to the DRC for its decision, with copies to the other Party and the Engineer.

Such reference shall state that it is given under this Sub-Clause.

For a DRC of six persons, the DRC shall be deemed to have received such reference on the date when it is received by thechairman of the DRC.

Three Parties shall promptly make available to the DRC all such additional information, further access to the Site, and appropriate facilities, as the DRC may require for the purposes of making a decision on such dispute.

The DRC shall be deemed to be not acting as arbitrator(s).

Within 7 days after receiving such reference, or within such other period as may be proposed by the DRC and approved by bothParties, the DRC shall give its decision, which shall be reasoned and shall state that it is given under this Sub-Clause.

The decision shall be binding on both Parties, who shall promptly give effect to it unless and until it shall be revised in an amicable settlement or an arbitral award as described below.

Unless the Contract has already been abandoned, repudiated or terminated, the Contractor shall continue to proceed with the Works in accordance with the Contract.

If either Party is dissatisfied with the DRC's decision, then either Party may, within 28 days after receiving the decision, give notice to the other Party of it dis-satisfaction.

If the DRC fails to give its decision within the period of 7 days (oras otherwise approved) after receiving such reference, then either Party may, within 7 days after this period has expired, give noticeto the other Party of it dis-satisfaction.

In either event, this notice of dis-satisfaction shall state that it is given under this Sub Clause and shall set out the matter in dispute and the reason(s) for dis-satisfaction.

Except as stated in Sub-Clause 20.7 [Failure to Comply with Dispute Resolution Committee's Decision] and Sub-Clause 20.8 [Expiry of Dispute Adjudication Board's Appointment], neither Party shall be entitled to commence arbitration of a dispute unless a notice of dissatisfaction has been given in accordance with this Sub-Clause.

If the DRC has given its decision as to a matter in dispute to both Parties, and no notice of dis-satisfaction has been given by either Party within 7 days after it received the DRC's decision, then the decision shall become final and binding upon both Parties.

20.5 Amicable Settlement

Where notice of dis-satisfaction has been given under Sub-Clause 20.4 above, both Parties shall attempt to settle the dispute amicably before the commencement of arbitration.

However, unless both Parties agree otherwise, arbitration may be commenced on or after the fifty-sixth day after the day on which notice of dis-satisfaction was given, even if no attempt at amicable settlement has been made.

20.6 Arbitration

Unless settled amicably, any dispute in respect of which the DRC's decision (if any) has not become final and binding shall be finally settled by international arbitration.

Unless otherwise agreed by both Parties:

a) the dispute shall be finally settled under the Rules of Arbitration of the International Chamber of Commerce,

b) the dispute shall be settled by three arbitrators appointed in accordance with these Rules, and

c) the arbitration shall be conducted in the language for communications defined in Sub-Clause 1.4 [Law and Language].

The arbitrator(s) shall have full power to open up, review and revise any certificate, determination, instruction, opinion or valuation of the Engineer, and any decision of the DRC, relevant to the dispute.

Nothing shall disqualify the Engineer from being called as a witness and giving evidence before the arbitrator(s) on any matterwhatsoever relevant to the dispute.

Neither Party shall be limited in the proceedings before the arbitrator(s) to the evidence or arguments previously put beforethe DRC to obtain its decision, or to the reasons for dis-satisfaction given in its notice of dis-satisfaction.

Any decision of the DRC shall be admissible in evidence in the arbitration.

Arbitration may be commenced prior to or after completion of the Works.

The obligations of the Parties, the Engineer and the DRC shall notbe altered by reason of any arbitration being conducted during the progress of the Works.

20.7 Failure to Comply with Dispute Resolution Committee Decision

In the event that

a) neither Party has given notice of dis-satisfaction within theperiod stated in Sub Clause 20.4 [Obtaining Dispute Resolution Committee Decision],

b) the DRC's related decision (if any) has become final and binding, and

c) a Party fails to comply with this decision, then the other Partymay, without prejudice to any other rights it may have, refer the failure itself to arbitration under Sub-Clause 20.6 [Arbitration]. Sub-Clause 20.4 [Obtaining Dispute ResolutionCommittee Decision] and Sub-Clause 20.5 [Amicable Settlement] shall not apply to this reference.

20.8 Expiry of Dispute Resolution Committee Appointment

If a dispute arises between the Parties in connection with, or arising out of, the Contract or the execution of the Works and there is no DRC in place, whether by reason of the expiry of the DRC's appointment or otherwise:

a) Sub-Clause 20.4 [Obtaining Dispute Resolution Committee Decision] and Sub Clause 20.5 [Amicable Settlement] shallnot apply, and
b) the dispute may be referred directly to arbitration under Sub-Clause 20.6 [Arbitration].

Maintenance

Maintenance during Construction Period

i. During the Construction Period, the Contractor shall maintain,at its cost, the existing structure of the Designed Work so thatthe traffic worthiness and safety thereof are at no time materially inferior as compared to their condition on Appointed Date, and shall undertake the necessary repair andmaintenance works for this purpose; provided that the Contractor may, at its cost, interrupt and divert the flow of traffic if such interruption and diversion is necessary for the efficient progress of Works and conforms to Good Industry Practice; provided further that such interruption and diversion shall be undertaken by the Contractor only with the prior written approval of the Engineer which approval shall not be unreasonably withheld.

For the avoidance of doubt, it is agreed that the Contractor shall at all times be responsible for ensuring safe operation of the WORK.

It is further agreed that in the event the Work includes construction of a bypass or tunnel and realignment of the existingcarriageway, the Contractor shall maintain the existing Work in such sections until the new Works are open to the public.

ii. Notwithstanding anything to the contrary contained in this Agreement, in the event of default by the Contractor in discharging the obligations specified in Clause 10.4 (I) above,the Engineer shall get these maintenance works completed in the manner recommended by the Engineer to avoid public inconvenience at the risk and cost of the Contractor in order to keep the road in traffic worthy condition.

Maintenance Manual No later than 60 (sixty) days prior to theProject Completion Date, the Contractor shall, in consultation with the Engineer's, evolve a maintenance manual (the "Maintenance Manual") for the regular and preventive maintenance of the Project Highway in conformity with the Specifications and Standards, safety requirements and Good Industry Practice, and shall provide 5 (five) copies thereof to the Engineer's. The Engineer shall review the Maintenance Manual within 15 (fifteen) days of its receipt and communicate its comments to the Contractor for necessary modifications, if any.

21.1 Maintenance obligations of the Contractor

(I) The Contractor shall maintain the Work for a period of 100 years or till dismantling of the structure, corresponding to the Defects Liability Period, commencing from the date of the Completion Certificate (the "Maintenance Period").

For the performance of its Maintenance obligations, the Contractor shall be paid:

(a) For newly developed Infrastructures with 100 years Maintenance Period or more.

Maintenance cost should be calculated as per life condition year wise and on cost plus basis.

During the Maintenance Period, the Authority shall provide to the Contractor access to the Site for Maintenance in accordance withthis Agreement.

The obligations of the Contractor hereunder shall include safe and smooth maintenance.

In respect of any Defect or deficiency not specified in Schedule-E, the

Contractor shall, at its own cost, undertake repair or rectification in accordance with Good Industry Practice, save and except to the extent that such Defect or deficiency shall have arisen on account of any willful default or neglect of the Authority or a Force Majeure Event.

21.2 Maintenance Requirements-

The Contractor shall ensure and procure that at all times during the Maintenance Period, conforms to the maintenancerequirements set forth in Schedule-E (the "Maintenance Requirements").

21.3 Maintenance Program me

(i) The Contractor shall prepare a monthly maintenance program me (the "Maintenance Program me") in consultation withthe Engineer and submit the same to the Engineer not later than 10 (ten) days prior to the commencement of the month in which the Maintenance is to be carried out.

For this purpose, a joint monthly inspection by the Contractor andthe Engineer shall be undertaken.

The Maintenance Program me shall contain the following:

a) The condition of the work in the format prescribed by theEngineer;

b) the proposed maintenance Works; and

c) deployment of resources for maintenance Works.

21.4 Safety, vehicle breakdowns and accidents

i. The Contractor shall ensure safe conditions for the Users, andin the event of unsafe conditions, lane closures, diversions, vehicle breakdowns and accidents, it shall follow the relevant operating procedures for removal of obstruction and debris without delay.

Such procedures shall conform to the provisions of this Agreement, Applicable Laws, Applicable Permits and Good Industry Practice.

ii. The Contractor shall maintain and operate a round-the-clock vehicle rescue post.

21.52

1.5 Lane closure

I. The Contractor shall not close any lane of the Work for undertaking maintenance works except with the prior written approval of the Engineer.

21.6

2

1.6 Reduction of payment for non-performance of Maintenance obligations

i. In the event that the Contractor fails to repair or rectify any Defect or deficiency set forth in Schedule-E within the period specified therein, it shall be deemed as failure of performance of Maintenance obligations by the Contractor and the Engineer shall be entitled to effect reduction in monthly lump sum payment for maintenance, without prejudice to the rights of the Authority under this Agreement, including Termination thereof.

ii. If the nature and extent of any Defect justifies more time forits repair or rectification than the time specified in Schedule-E, the Contractor shall be entitled to additional time in conformity with Good Industry Practice.

Such additional time shall be determined by the Engineer's and conveyed to the Contractor and the Engineer with reasons thereof.

21.7 Engineer's right to take remedial measures in the event the Contractor does not maintain and/or repair the Work or any part thereof in conformity with the Maintenance Requirements, the Maintenance Manual or the Maintenance Program me, as the case may be, and fails to commence remedial works within 15 (fifteen) days of receipt of the Maintenance Inspection Report under Clause

15.2 or a notice in this behalf from the Engineer, as the case may be, the Engineer shall, without prejudice to its rights under this Agreement including Termination thereof, be entitled to undertake such remedial measures at the cost of the Contractor, and to recover its cost from the Contractor.

In addition to recovery of the aforesaid cost, a sum equal to 20% (twenty per cent) of such cost shall be paid by the Contractor to the Engineer as Damages.

21.8 Restoration of loss or damage to Project Save and except as otherwise expressly provided in this Agreement, in the event that the Project or any part thereof suffers any loss or damage during the Maintenance from any cause attributable to the Contractor, the Contractor shall, at its cost and expense, rectify and remedy such loss or damage forthwith so that the Project conforms to the provisions of this Agreement.

21.9 Overriding powers of the Engineer:

i. If in the reasonable opinion of the Engineer, the Contractor isin material breach of its obligations under this Agreement and, in particular, the Maintenance Requirements, and such breach is causing or likely to cause material hardship or danger to the Users and pedestrians, the Engineer may, without prejudice to any of its rights under this Agreement including Termination thereof, by notice require the Contractor to take reasonablemeasures immediately for rectifying or removing such hardship or danger, as the case may be.

ii. In the event that the Contractor, upon notice under Clause 14.9 (I), fails to rectify or remove any hardship or danger within a reasonable period, the Engineer may exercise overriding powers under this Clause 14.9 (ii) and take over the performance of any or all the obligations of the Contractor tothe extent deemed necessary by it for rectifying or removing such hardship or danger; provided that the exercise of such overriding powers by the Engineer shall be of no greater scope and of no longer 80 duration than is reasonably required hereunder; provided further that any costs and expenses incurred by the Engineer in discharge of its obligations hereunder shall be recovered by the Engineer from the Contractor, and the Engineer shall be entitled to deduct any such costs and expenses incurred from the payments due to the Contractor under Clause 19.7 for the performance of its Maintenance obligations.

iii. In the event of a national emergency, civil commotion or any other circumstances specified in Clause 21.3, the Engineer may take over the performance of any or all the obligations of the Contractor to the extent deemed necessary by it, and exercise such control over the Project or give such directions to the Contractor as may be deemed necessary; provided that the exercise of such overriding powers by the shall be of no greater scope and of no longer duration than is reasonably required in the circumstances which caused the exercise ofsuch overriding power by the Engineer.

For the avoidance of doubt, it is agreed that the consequences of such action shall be dealt in accordance with the provisions ofArticle 21.

It is also agreed that the Contractor shall comply with such instructions as the Engineer may issue in pursuance of the provisions of this Clause

21.9 (iii), and shall provide assistance and cooperation to the Authority, on a best effort basis, for performance of its obligations hereunder.

21.10

1.10 Taking over Certificate

The Maintenance Requirements set forth in Schedule-E having been duly carried out, for Maintenance Period as set forth in Clause

21.10 (I) having been expired and Engineer determining the Testson Completion of Maintenance to be successful in accordance with Schedule-Q, the Engineer will issue Taking Over Certificate to the Contractor substantially in the format set forth in Schedule-R.

Supervision and Monitoring duringMaintenance

22.1 Inspection by the Contractor

i. The Engineer's shall undertake regular inspections to evaluate continuously the compliance with the Maintenance Requirements.

ii. The Contractor shall carry out a detailed pre-monsoon inspection of all bridges, culverts and drainage systems in accordance with the guidelines prepared by the Engineer or Engineer Representative.

Report of this inspection together with details of proposedmaintenance works as required shall be conveyed to the Engineer's forthwith.

The Contractor shall complete the proposed maintenance works before the onset of the monsoon and send a compliance report to the Engineer's.

Post monsoon inspection shall be undertaken by the Contractor and the inspection report together with details of any damages observed and proposed action to remedy the same shall be conveyed to the Engineer's forthwith.

22.2 Inspection and payments

i. The Engineer's may inspect the Project at any time, but at least once every month, to ensure compliance with the Maintenance Requirements.

It shall make a report of such inspection ("Maintenance Inspection Report") stating in reasonable detail the Defects or deficiencies, if any, with particular reference to the Maintenance Requirements, the Maintenance Manual, and the Maintenance Program me, andsend a copy thereof to the Engineer and the Contractor within 10 (ten) days of such inspection.

ii. After the Contractor submits to the Engineer the Monthly Maintenance Statement for the Project pursuant to Clause 19.6, the Engineer shall carry out an inspection within 10 (ten)days to certify the amount payable to the Contractor.

The Engineer shall inform the Contractor of its intention to carry out the inspection at least 3 (three) business days in advance of such inspection. The Contractor shall assist the Engineer in verifying compliance with the Maintenance Requirements.

iii. For each case of non-compliance of Maintenance Requirementsas specified in the inspection report of the Engineer's, the Engineer's shall calculate the amount of reduction in paymentin accordance with the formula specified in Schedule-M.

iv. Any deduction made on account of non-compliance will not be paid subsequently even after establishing the compliance thereof.

Such deductions will continue to be made every month until the compliance is procured.

22.3 Tests For determining that the Project conforms to the Maintenance Requirements, the Engineer's shall require the Contractor to carry out, or cause to be carried out, Tests specified by it in accordance with Good Industry Practice.

The Contractor shall, with due diligence, carry out or cause to be carried out all such Tests in accordance with the instructions of the Engineer's and furnish the results of such Tests forthwith to the Engineer.

At any time during the Maintenance Period, the Engineer may appoint an external technical auditor to conduct an audit of the quality of the Works.

The Auditor in the presence of the representatives of the Contractor and the Engineer's shall carry out the Tests and/ or collect samples for testing in the laboratory.

The timing, the testing equipment and the sample size of this audit shall be as decided by the Engineer.

The findings of the audit, to the extent accepted by the, shall be Engineer notified to the Contractor and the Engineer's for taking remedial measures.

After completion of the remedial measures by the Contractor, the auditor shall undertake a closure audit and this process will continue till the remedial measures have brought the maintenanceworks into compliance with the Specifications and Standards.

The Contractor shall provide all assistance as may be required bythe auditor in the conduct of its audit hereunder. Notwithstanding anything contained in this Clause 15.3, the external technical audit shall not affect any obligations of the Contractor or the Engineer's under this Agreement.

22.4 Reports of unusual occurrence The Contractor shall, during the Maintenance Period, prior to the close of each day, send to the Engineer's, by facsimile or e- mail, a report stating accidents and unusual occurrences on the Project relating to the safety and security of the Users and Project:

A monthly summary of such reports shall also be sent within 3 (three) business days of the closing of month. For the purposes of this Clause 15.4, accidents and unusual occurrences on the Project shall include:

a. accident, death or severe injury to any person.

b. damaged or dislodged fixed equipment.

c. flooding of Project; and

d. any other unusual occurrence

APPENDIX

General Conditions of Dispute Adjudication Agreement

1. Definitions Each "Dispute

Adjudication Agreement" is a tripartite agreement by and between:

a. the "Employer (Authority & Engineer)".

b. the "Contractor"; and

c. the "Member" who is defined in the Dispute Adjudication Agreement as being:

i. the sole member of the "DRC" (or "adjudicator") and, wherethis is the case, all references to the "Other Members" do not apply, or

ii. one of the six persons who are jointly called the "DRC" (or "dispute resolution committee") and, where this is the case, the other persons are called the "Other Members".

The Engineer and the Contractor have entered (or intend to enter) into a contract, which is called the "Contract" and is defined in the Dispute Adjudication Agreement, which incorporates this Appendix.

In the Dispute Adjudication Agreement, words and expressions which are not otherwise defined shall have the meanings assignedto them in the Contract.

2. General Provisions

Unless otherwise stated in the Dispute Adjudication Agreement, it shall take effect on the latest of the following dates:

a. the Commencement Date defined in the Contract,

b. when the Engineer, the Contractor and the Member have each signed the Dispute Adjudication Agreement, or

c. when the Engineer, the Contractor and each of the Other Members (if any) have respectively each signed a dispute adjudication agreement.

When the Dispute Adjudication Agreement has taken effect, the Engineer and the Contractor shall each give notice to the Member accordingly. If the Member does not receive either notice within six months after entering into the Dispute Adjudication Agreement, it shall bevoid and ineffective.

This employment of the Member is a permanent or personal appointment. At any time, the Member may give not less than 70days' notice of resignation to the Engineer and to the Contractor, and the Dispute Adjudication Agreement shall terminate upon theexpiry of this period. No assignment or subcontracting of the Dispute Adjudication Agreement is permitted without the prior written agreement of allthe parties to it and of the Other Members (if any).

3. Warranties

The Member warrants and agrees that he/she is and shall be impartial and independent of the Employer, the Contractor and theEngineer.

The Member shall promptly disclose, to each of them and to the Other Members (if any), any fact or circumstance which might appear inconsistent with his/her warranty and agreement of impartiality and independence.

When appointing the Member, the Engineer and the Contractor relied upon the Member's representations that he/she is:

a. experienced in the work which the Contractor is to carry outunder the Contract,
b. experienced in the interpretation of contract documentation,and
c. fluent in the language for communications defined in theContract.

4. General Obligations of the Member

The Member shall:

a) have no interest financial or otherwise in the Engineer, theContractor or the Engineer, nor any financial interest in the Contract except for payment under the Dispute Adjudication Agreement.

b) not previously have been employed as a consultant or otherwise by the Engineer, the Contractor or the Engineer, except in such circumstances as were disclosed in writing to the Engineer and the Contractor before they signed the Dispute Adjudication Agreement;

c) have disclosed in writing to the Engineer, the Contractor andthe Other Members (if any), before entering into the Dispute Adjudication Agreement and to his/her best knowledge and recollection, any professional or personal relationships with any director, officer or employee of the Engineer, the Contractor or the Engineer, and any previous involvement inthe overall project of which the Contract forms a part;

d) not, for the duration of the Dispute Adjudication Agreement, be employed as a consultant or otherwise by the Engineer, the Contractor or the Engineer, except as may be agreed in writing by the Engineer, the Contractor and the Other Members (if any);

e) comply with the annexed procedural rules and with Sub- Clause 20.4 of the Conditions of Contract;

f) not give advice to the Engineer, the Contractor, the Engineer's Personnel or the Contractor's Personnel concerning the conduct of the Contract, other than in accordance with the annexed procedural rules;

g) not while a member enters into discussions or make any agreement with the Engineer, the Contractor or the Engineer regarding employment by any of them, whether as a consultant or otherwise, after ceasing to act under the Dispute Adjudication Agreement;

h) ensure his/her availability for all site visits and hearings as are necessary;

i) become conversant with the Contract and with the progress of the Works (and of any other parts of the project of which the Contract forms a part) by studying all documents received which shall be maintained in a current working file;

j) treat the details of the Contract and all the DRC's activities and hearings as private and confidential, and not publish or disclose them without the prior written consent of the Engineer, the Contractor and the Other Members (if any); and

k) be available to give advice and opinions, on any matterrelevant to the Contract when requested by both the Engineer and the Contractor, subject to the agreement of the Other Members (if any).

5. General Obligations of the Engineer and the Contractor

The Engineer, the Contractor, the Engineer's Personnel and the Contractor's Personnel shall not request advice from or consultation with the Member regarding the Contract, otherwise than in the normal course of the DAB's activities under theContract and the Dispute Adjudication Agreement, and except tothe extent that prior agreement is given by the Engineer, the Contractor and the Other Members (if any).

The Engineer and the Contractor shall be responsible for compliance with this provision, by the Engineer's Personnel and the Contractor's Personnel respectively. The Engineer and the Contractor undertake to each other and to the Member that the Member shall not, except as otherwise agreed in writing by the Engineer, the Contractor, the Member and the Other Members (ifany):

a. be appointed as an arbitrator in any arbitration under the Contract;
b. be called as a witness to give evidence concerning any dispute before arbitrator(s) appointed for any arbitration under the Contract; or
c. be liable for any claims for anything done or omitted in the discharge or purported discharge of the Member's functions, unless the act or omission is shown to have been in bad faith.

The Engineer and the Contractor hereby jointly and severally indemnify and hold the Member harmless against and from claims

from which he is relieved from liability under the preceding paragraph.

Whenever the Engineer or the Contractor refers a dispute to the DRC under Sub Clause 20.4 of the

Conditions of Contract, which will require the Member to make asite visit and attend a hearing, the Employer or the Contractor shall provide appropriate security for a sum equivalent to the reasonable expenses to be incurred by the Member.

No account shall be taken of any other payments due or paid to the Member.

6. Payment

The Member shall be paid as follows, in the currency named in the Dispute Adjudication Agreement:

(a) a retainer fee per calendar month, which shall be considered

as payment in full for:

i. being available on 28 days' notice for all site visits and hearings;

ii. becoming and remaining conversant with all project developments and maintaining relevant files;

iii. all office and overhead expenses including secretarial services, photocopying and office supplies incurred in connection withhis duties; and

iv. all services performed hereunder except those referred to in subparagraphs (b) and (c) of this Clause.

The retainer fee shall be paid with effect from the last day of the calendar month in which the Dispute Adjudication Agreement becomes effective; until the last day of the calendar month in which the Taking-Over Certificate is issued for the whole of the Works.

With effect from the first day of the calendar month following the month in which the Taking-Over Certificate is issued for the whole of the Works, the retainer fee shall be reduced by 50%.

This reduced fee shall be paid until the first day of the calendar month in which the Member resigns or the Dispute Adjudication Agreement is otherwise terminated.

(b) a daily fee which shall be considered as payment in full for:

i. each day or part of a day up to a maximum of two days' travel time in each direction for the journey between the Member's home and the site, or another location of a meeting with the Other Members (if any);

ii. each working day on site visits, hearings or preparing decisions; and

iii. each day spent reading submissions in preparation for a hearing.

c. all reasonable expenses incurred in connection with theMember's duties, including the cost of telephone calls, courier charges, faxes and telexes, travel expenses, hotel and subsistence costs: a receipt shall be required for each item in excess of five percent of the daily fee referred to in sub- paragraph (b) of this Clause;

d. any taxes properly levied in the Country on payments made tothe Member (unless a national or permanent resident of the Country) under this Clause 6.

The retainer and daily fees shall be as specified in the Dispute Adjudication Agreement.

Unless it specifies otherwise, these fees shall remain fixed for the first 24 calendar months and shall thereafter be adjusted by agreement between the Employer, the Contractor and the Member,at each anniversary of the date on which the Dispute Adjudication Agreement became effective.

The Member shall submit invoices for payment of the monthly retainer and air fares quarterly in advance.

Invoices for other expenses and for daily fees shall be submitted following the conclusion of a site visit or hearing.

All invoices shall be accompanied by a brief description of activities performed during the relevant period and shall be addressed to the Contractor.

The Contractor shall pay each of the Member's invoices in full within 56 calendar days after receiving each invoice and shall apply to the Engineer (in the Statements under the Contract) for reimbursement of one-half of the amounts of these invoices.

The Engineer shall then pay the Contractor in accordance with the Contract.

If the Contractor fails to pay to the Member the amount to which he/she is entitled under the Dispute Adjudication Agreement, the Engineer shall pay the amount due to the Member and any other amount which may be required to maintain the operation of the DRC; and without prejudice to the Engineer's rights or remedies.

In addition to all other rights arising from this default, the Engineer shall be entitled to reimbursement of all sums paid in excess of one-half of these payments, plus all costs of recoveringthese sums and financing charges calculated at the rate specified inSub-Clause 14.8 of the Conditions of Contract.

If the Member does not receive payment of the amount due within70 days after submitting a valid invoice, the Member may:

i. suspend his/her services (without notice) until the payment is received, and/or

ii. resign his/her appointment by giving notice under Clause 7.

7. Termination

At any time:

i. the Engineer and the Contractor may jointly terminate the Dispute Adjudication Agreement by giving 42 days' notice tothe Member; or

ii. the Member may resign as provided for in Clause 2. If the Member fails to comply with the Dispute Adjudication Agreement, the Engineer and the contractor may, without prejudice to their other rights, terminate it by notice to the Member.

The notice shall take effect when received by the Member.

If the Engineer or the Contractor fails to comply with the Dispute Adjudication Agreement, the Member may, without prejudice to his other rights, terminate it by notice to the Engineer and the Contractor.

The notice shall take effect when received by them both. Any such notice, resignation and termination shall be final and binding on the Engineer, the Contractor and the Member.

However, a notice by the Engineer or the Contractor, but not by both, shall be of no effect.

8. Default of the Member

If the Member fails to comply with any obligation under Clause 4, he/she shall not be entitled to any fees or expenses hereunder and shall, without prejudice to their other rights, reimburse each of the Engineer and the Contractor for any fees and expenses received by the Member and the Other Members (if any), for proceedings or decisions (if any) of the DRC which are rendered void or ineffective.

9. Disputes

Any dispute or claim arising out of or in connection with this Dispute Adjudication Agreement, or the breach, termination or invalidity thereof, shall be finally settled under the Rules of Arbitration of the International Chamber of Commerce by one arbitrator appointed in accordance with these Rules of Arbitration.

Annex Procedural Rules

1. Unless otherwise agreed by the Engineer and the Contractor,the DRC shall regular visit the site for 7 days at intervals of 90days.
2. The timing of and agenda for each site visit shall be as agreed jointly by the DRC, the Engineer and the Contractor, or in the absence of agreement, shall be decided by the DRC. The purpose of site visits is to enable the DRC to become and remain acquainted with the progress of the Works and of any actual or potential problems or claims.
3. Site visits shall be attended by the Engineer, the Contractor and the Engineer and shall be coordinated by the Engineer in co-operation with the Contractor. The Engineer shall ensure the provision of appropriate conference facilities and secretarial and copying services. At the conclusion of each site visit andbefore leaving the site, the DRC shall prepare a report on its activities during the visit and shall send copies to the Engineer and the Contractor.
4. The Engineer and the Contractor shall furnish to the DRC one copy of all documents which the DRC may request, including Contract documents, progress reports, variation instructions, certificates and other documents pertinent to the performanceof the Contract. All communications between the DRC and the Engineer or the Contractor shall be copied to the other Party. If the DRC comprises Six persons, the Employer and the Contractor shall send copies of these requested documents and these communications to each of these persons.

5. If any dispute is referred to the DRC in accordance with Sub- Clause 20.4 of the Conditions of Contract, the DRC shallproceed in accordance with Sub-Clause 20.4 and these Rules. Subject to the time allowed to give notice of a decision and other relevant factors, the DRC shall:

a. act fairly and impartially as between the Employer and the Contractor, giving each of them a reasonable opportunity of putting his case and responding to the other's case, and

b. adopt procedures suitable to the dispute, avoiding unnecessarydelay or expense.

6. The DRC may conduct a hearing on the dispute, in which event it will decide on the date and place for the hearing and may request that written documentation and arguments from the Engineer and the Contractor be presented to it prior to or at the hearing.

7. Except as otherwise agreed in writing by the Engineer and the Contractor, the DRC shall have power to adopt an inquisitorial procedure, to refuse admission to hearings or audience at hearings to any persons other than representatives of the Engineer, the Contractor and the Engineer, and to proceed inthe absence of any party who the DRC is satisfied received notice of the hearing; but shall have discretion to decide whether and to what extent this power may be exercised.

8. The Engineer and the Contractor empower the DRC, among other things, to:

a. establish the procedure to be applied in deciding a dispute,

b. decide upon the DRC's own jurisdiction, and as to the scope of any dispute referred to it,

c. conduct any hearing as it thinks fit, not being bound by any rules or procedures other than those contained in the Contractand these Rules,

d. take the initiative in ascertaining the facts and matters required for a decision,

e. make use of its own specialist knowledge, if any,

f. decide upon the payment of financing charges in accordance with the Contract,

g. decide upon any provisional relief such as interim or conservatory measures, and

h. open up, review and revise any certificate, decision, determination, instruction, opinion or valuation of the Engineer, relevant to the dispute.

9. The DRC shall not express any opinions during any hearing concerning the merits of any arguments advanced by the Parties.

Thereafter, the DRC shall make and give its decision in accordance with Sub-Clause 20.4, or as otherwise agreed by the Engineer and the Contractor in writing.

If the DRC comprises six persons:

a. it shall convene in private after a hearing, in order to have discussions and prepare its decision;

b. it shall endeavor to reach a unanimous decision: if this proves impossible the applicable decision shall be made by a majorityof the Members, who may require the minority Member to prepare a written report for submission to the Engineer and the Contractor; and

c. if a member fails to attend a meeting or hearing, or to fulfil any required function, the other Members may nevertheless proceed to make a decision, unless:

i. either the Engineer or the Contractor does not agree that theydo so, or

ii. the absent Member is the chairman, and he/she instructs the other Members not to make a decision.

GENERAL CONDITIONS

GUIDANCE FOR THE PREPARATION OF PARTICULAR CONDITIONS

FORMS OF LETTER OF TENDER, CONTRACT AGREEMENT AND DISPUTE ADJUDICATION AGREEMENT

CONDITIONS OF CONTRACT FOR CONSTRUCTION

FOR INFRASTRUCTURE DEVELOPMENT & MAINTENANCE WORK.

Guidance for the Preparation of Particular Conditions

Guidance for the Preparation of Particular Conditions

INTRODUCTION

The terms of the Conditions of Contract for Infrastructure Development Work and Maintenance have been prepared by the ENGINEER LIVE FOUNDATION (ELF) and are recommendedfor general use for the purpose of the Innovative Design, Research & Development, Bidding, construction and Maintenance of all types of Engineering works where tenders are invited on aninternational/National basis. Modifications to the Conditions may be required in some legal jurisdictions, particularly if they are to be used on domestic contracts.

The guidance hereafter is intended to assist writers of theParticular Conditions by giving options for various sub-clauses where appropriate. As far as possible, example wording is included, between lines. In some cases, however, only an aide-memoire is given.

Before incorporating any example wording, it must be checked to ensure that it is wholly suitable for the particular circumstances. Unless it is considered suitable, example wording should be amended before use.

Where example wording is amended, and in all cases where other amendments or additions are made, care must be taken to ensure thatno ambiguity is created, either with the General Conditions or between the clauses in the Particular Conditions.

In the preparation of the Conditions of Contract to be included in the tender documents for a contract, the following text can be used:

The Conditions of Contract comprise the "General Conditions", which form part of the "Conditions of Contract for Infrastructure Development Work

& Maintenance" First Edition 2023 publishedby the, Engineer Live Foundation (ELF) and the following "Particular Conditions", which include amendments and additions to such General Conditions.

There are no Sub-Clauses in the General Conditions which require data to be included in the Particular Conditions. As noted in sub- paragraph (ii) of the Foreword, the General Conditions refer to any necessary data being contained in the Appendix to Tender or (for technical matters) in the Specification.

ELF has to publish a document entitled "Tendering Procedure" which will presents a systematic approach to the selection of tenderers and the obtaining and evaluation of tenders. The document will intend to assist the Employer/Engineer to receive sound competitive tenders with a minimum of qualifications. ELF intends to update Tendering Procedure and to publish a guide to the use of these Conditions of Contract for Construction.

Notes on the Preparation of TenderDocuments

The tender documents should be prepared by suitably-qualified engineers who are familiar with the technical aspects of the requiredworks, and a review by suitably-qualified lawyers may be advisable.The tender documents issued to tenderers will consist of the Conditions of Contract, the Specification, the Drawings, and the Letter of Tender and Schedules for completion by the Tenderer. For this typeof contract, where the Works are valued by measurement, the Billof Quantities will usually be the most important Schedule.

A Daywork Schedule may also be necessary, to cover minor works to be evaluated at cost. In addition, each of the Tenderers should receive the data referred to in Sub-Clause 4.10, and the Instructions to Tenderers to advise them of any special matters which the Employer wishes them to take into account when pricing the Bill of Quantities but which are not to form part of the Contract. When the Engineer accepts the Letter of Tender, the Contract (which then comes into full force and effect) includes thesecompleted Schedules.

The Specification may include the matters referred to in some or all of the following Sub-Clauses:

1.8 Requirements for Contractor's Documents

1.13 Permissions being obtained by the Engineer

2.1 Phased possession of foundations, structures, plant or means ofaccess

4.1 Contractor's designs

4.6 Other contractors (and others) on the Site

4.7 Setting-out points, lines and levels of reference

4.14 Third parties

4.18 Environmental constraints

4.19 Electricity, water, gas and other services available on the Site

4.20 Engineer's Equipment and free-issue material

5.1 Nominated Subcontractors

6.6 Facilities for Personnel

7.2 Samples

7.4 Testing during manufacture and/or construction

9.1 Tests on Completion

13.5 Provisional Sums

Many Sub-Clauses in the General Conditions make reference to data being contained in the Appendix to Tender, providing a convenient location for the data which is usually required. The example form in this publication thus provides a check-list of thedata required; but there is no indication, either in the General Conditions or in the example Appendix to Tender, that this data is either prescribed by the Engineer or inserted by the Tenderer. TheEngineer should prepare the Appendix to Tender, based on this example form, with the elements completed to the extent of his requirements.

The Engineer may also require other data from Tenderers, and include a questionnaire in the Schedules.

The Instructions to Tenderers may need to specify any constraints on the completion of the Appendix to Tender and/or Schedules, and/or specify the extent of other information which each Tenderer is to include with his Tender. If each Tenderer is to produce a parent company guarantee and/or a tender security, these requirements (which apply prior to the Contract becoming effective) should be included in the Instructions to Tenderers: example forms are annexed to this documents Annexes A and B. The Instructions may include matters referred to in some or all of the following Sub-Clauses:

4.3 Contractor's Representative (name and curriculum vitae)

4.9 Quality Assurance system

9.1 Tests on Completion18 Insurances

1 Resolution of disputes

Clause 1 General Provisions

Sub-Clause 1.1 Definitions

It may be necessary to amend some of the definitions. For example:

1.1.3.1 the Base Date could be defined as a particular calendar date

1.1.4.6 one particular Foreign Currency may be required by the financing institution

1.1.4.8 a different currency may be required to be the contract Local Currency

1.1.6.2 the references to "Country" may be inappropriate for a cross-border Site

Sub-Clause 1.2 Interpretation

If the references to "profit" are to be more precisely specified, this Sub-Clause may be varied:

EXAMPLE At the end of Sub-Clause 1.2, insert:

In these Conditions, provisions including the expression "Cost plus reasonable profit" require this profit to be one-twentieth (5%) of this Cost.

Sub-Clause 1.5 Priority of Documents

An order of precedence is usually necessary, in case a conflict is subsequently found among the contract documents. If no order of precedence is to be prescribed, this Sub-Clause may be varied:

EXAMPLE Delete Sub-Clause 1.5 and substitute:

The documents forming the Contract are to be taken as mutually explanatory of one another. If an ambiguity or discrepancy is found, the priority shall be such as may be accorded by the governing law. The Engineer has authority to issue any instruction which he considers necessary to resolve an ambiguity or discrepancy.

Sub-Clause 1.6 Contract Agreement

The form of Agreement should be included in the tender documents as an annex to the Particular Conditions: an example form is includedat the end of this publication. If lengthy tender negotiations were necessary, it may be considered advisable for the ContractAgreement to record the Accepted Contract Amount, Base Date and/or Commencement Date. Entry into an Agreement may be necessary under applicable law.

Sub-Clause 1.14 Joint and Several Liability

For a major contract, detailed requirements for the joint venture may need to be specified. For example, it may be desirable for each member to produce a parent company guarantee: an example formis annexed to this document as Annex A.

These requirements, which apply prior to the Contract becoming effective, should be included in the Instructions to Tenderers. The Engineer will wish the leader of the joint venture to be appointedat an early stage, providing a single point of contact thereafter, and will not wish to be involved in a dispute between the members of a joint venture. The Engineer should scrutinize the joint venture agreement carefully, and it may have to be approved by the project's financing institutions.

Additional Sub-Clause Details to be Confidential

If confidentiality is required, an additional sub-clause may be added:

EXAMPLE SUB-CLAUSE

The Contractor shall treat the details of the Contract as private and confidential, except to the extent necessary to carry outobligations under it or to comply with applicable Laws. The Contractor shall not publish, permit to be published, or disclose any particulars of the Works in any trade or technical paper or elsewhere without the previous agreement of the Engineer.

Clause 2 The Employer

Employer appointing the Architect & Engineer and Authority.

Sub-Clause 2.1 Right of Access to the Site

If right of access cannot be granted, both early and thereafter exclusively, details should be given in the Specification.

Sub-Clause 2.3 Employer's Personnel

These provisions should be reflected in the Employer's contracts withany other contractors on the Site.

Clause 3 The Architect &Engineer & Authority

Sub-Clause 3.1 Architect & Engineer'sDuties and Authority

Any requirements for Employer's approval should be set out in the Particular Conditions:

EXAMPLE

The Authority & Architect & Engineer shall obtain the specific approval of the Employer before taking action under the following Sub-Clauses of these Conditions:

(a) Sub-Clause________________________________**

(b) Sub-Clause________________________________**

(insert number; describe action, unless all require approval)

This list should be extended or reduced as necessary. If the obligationto obtain the approval of the Employer only applies beyond certain limits, financial or otherwise, the example wording should be varied.

Additional Sub-Clause Management MeetingsEXAMPLE SUB-CLAUSE The Engineer or the Contractor's Representative may require the other to attend a management meeting in order to review the arrangements for future work. The Engineer shall record thebusiness of management meetings and supply copies of the record tothose attending the meeting and to the Engineer. In the record, responsibilities for any actions tobe taken shall be in accordance with the Contract.

Clause 4 The Contractor

Sub-Clause 4.1Contractor's GeneralObligations

Occasionally, there may be an item of Temporary Works for which the Contractor will not be fully responsible. For example, the Contract may specify temporary arrangements for river diversion which have been designed by the Engineer. In these cases, the Sub- Clause may require amendment, taking account of the type of this item of Temporary Works, and of the extent of the Engineer's responsibility.

Sub-Clause 4.2 Performance Security

The acceptable form(s) of Performance Security should be included in the tender documents, annexed to the Particular Conditions. Example forms are annexed to this document as Annex C and Annex D. They incorporate two sets of Uniform Rules published by the International Chamber of Commerce (the "ICC", which is basedat 38 Cours Albert 1er, 75008 Paris, France), which also publishes guides to these Uniform Rules. These example forms and the wording of the Sub-Clause may have to be amended to comply with applicable law.

EXAMPLE At the end of the second paragraph of Sub-Clause 4.2, insert:

If the Performance Security is in the form of a bank guarantee, it shall be issued either (a) by a bank located in the Country, or (b) directly by a foreign bank acceptable to the Engineer. If the Performance Security is not in the form of a bank guarantee, it shall be furnished by a

financial entity registered, or licensed to do business, in the Country.

Sub-Clause 4.3 Contractor's Representative

If the Representative is known at the time of submission of the Tender, the Tenderer may propose the Representative. The Tenderer may wish to propose alternatives, especially if the contract award seems likely to be delayed. If the ruling language is not the same as the language for day-to-day communications (under Sub-Clause 1.4),or if for any other reason it is necessary to stipulate that the Contractor's Representative shall be fluent in a particular language, one of the following sentences may be added.

EXAMPLE At the end of Sub-Clause 4.3, add:

The Contractor's Representative and all these persons shall alsobe fluent in

________________________________(insert name of language)

EXAMPLE At the end of Sub-Clause 4.3, add:

If the Contractor's Representative, or these persons, is not fluentin ________________________________(Insert name of language), the Contractor shall make a competent interpreter available during all working hours.

Sub-Clause 4.4Subcontractors

The wording in the General Conditions includes the conditions which will usually be applicable. If less (or no) consent is required, some (or all) of sub-paragraphs (a) to (d) may be deleted, or qualified in the **Particular Conditions**

EXAMPLE Prior consent shall not be required if the value of the subcontract is less than 0.01% of the Accepted Contract Amount.

A sentence may be added to increase the extent to which consent is required:

EXAMPLE

The prior consent of the Engineer shall be obtained to thesuppliers of the following Materials:

(Insert details: for example, specific manufactured or prefabricated items)

A sentence may be added in order to encourage the Contractor touse local contractors:

EXAMPLE

Where practicable, the Contractor shall give a fair and reasonable opportunity for contractors from the Country to be appointed as Sub-contractors.

Sub-Clause 4.8 Safety Procedures

If the Contractor is sharing occupation of the Site with others, itmay not be appropriate for him to provide some of the listed items.In these circumstances, the Engineer's obligations should be specified.

Sub-Clause 4.9 Quality Assurance

The wording in the General Conditions imposes the requirement ofa quality assurance system in accordance with details specified inthe Contract. If inappropriate, this Sub-Clause may be deleted.

Sub-Clause 4.12 Unforeseeable Physical Conditions

In the case of major sub-surface works, the allocation of the risk of sub-surface conditions is an aspect which should be considered whentender documents are being prepared. If this risk is to be shared between the parties, the Sub-Clause may be amended:

EXAMPLE Delete sub-paragraph (b) of Sub-Clause 4.12 and substitute:

(b) ______________________________ payment for any such

Cost, ______________________________ percent

(____%) of which shall be included in the Contract Price (the balance percent of the Cost shall be borne by the Contractor).

Sub-Clause 4.17 Contractor's Equipment

If the Contractor is not to provide all the Contractor's Equipment necessary to complete the Works, the Engineer's obligations should be specified: see Sub-Clause 4.20. If vesting of Contractor's Equipment is required, further paragraphs may be added, subject to their being consistent with applicable laws:

EXAMPLE At the end of Sub-Clause 4.17, add the following paragraphs:

Contractor's Equipment which is owned by the Contractor (either directly or indirectly) shall be deemed to be the property of the Engineer with effect from its arrival on the Site. This vesting of property shall not:

(a) effect the responsibility or liability of the Engineer,

(b) prejudice the right of the Contractor to the sole use of the vested Contractor's Equipment for the purpose of the Works, or

(c) affect the Contractor's responsibility to operate and maintain Contractor's Equipment.

The property in each item shall be deemed to revest in the Contractor when he is entitled either to remove it from the Site orto receive the Taking-Over Certificate for the Works, whichever occurs first.

Sub-Clause 4.19 Electricity, Water and Gas

If services are to be available for the Contractor to use, the Specification should give details, including locations and prices.

Sub-Clause 4.20 Authority's Equipment and Free-IssueMaterial

For this Sub-Clause to apply, the Specification should describe each item which the Authority will provide and/or operate and should specify all necessary details. With some types of facilities, further provisions may be necessary, in order to clarify aspects such as liability and insurance.

Sub-Clause 4.22 Security of the Site

If the Contractor is sharing occupation of the Site with others, it may not be appropriate for him to be responsible for its security. In these circumstances, the Engineer's obligations should be specified.

Clause5 Nominated Subcontractors

In most cases under Sub-Clause 4.4, the Contractor selects Subcontractors, subject to any constraints specified in the Contract. Clause 5 provides for the particular situation whereby the Engineer may select a Subcontractor, although the second sentence of Sub- Clause 4.4 should still apply.

The sub-paragraphs of Sub-Clause 5.2 indicate some of the problems which may have to be overcome.

If a nominated Subcontractor is to be required, full details shouldbe included in the tender documents. If the Engineer anticipatesthat a Subcontractor is to be instructed under Clause 13but is not to be a nominated Subcontractor, Clause 5 should be amended, describing the particular circumstances.

Clause 6 Staff and Labor

Sub-Clause 6.5 Working Hours

If the Engineer does not wish to specify working hours in the Appendix to Tender, or to restrict them to the times specified by the Tenderer (in order to plan the Engineer's supervision, for example), this Sub-Clause may be deleted.

Sub-Clause 6.6 Facilities for Staff and Labor

If the Engineer will make some accommodation available, his obligations to do so should be specified.

Sub-Clause 6.8 Contractor's Superintendence

If the ruling language is not the same as the language for day-to-day communications (under Sub- Clause 1.4), or if for any other reason itis necessary to stipulate that the Contractor's superintending staff shall be fluent in a particular language, the following sentence may be added.

EXAMPLE

Insert at the end of Sub-Clause 6.8:

A reasonable proportion of the Contractor's superintending staff shall have a working knowledge of (Insert name of language), or the Contractor shall have a sufficient number of competent interpreters available on Site during all working hours.

Additional Sub-Clauses

It may be necessary to add a few sub-clauses to take account of the circumstances and locality of the Site:

EXAMPLE SUB-CLAUSE

Foreign Staff and Labor

The Contractor may import any personnel who are necessary for the execution of the Works. The Contractor must ensure that these personnel are provided with the required residence visas and work permits. The Contractor shall be responsible for the return to the place where they were recruited or to their domicile of imported Contractor's Personnel. In the event of the death in the Country ofany of these personnel or members of their families, the Contractor shall similarly be responsible for making the appropriate arrangements for their return or burial.

EXAMPLE SUB-CLAUSE

Measures against Insect and Pest Nuisance

The Contractor shall at all times take the necessary precautions to protect all staff and labor employed on the Site from insect and pest nuisance, and to reduce their danger to health. The Contractor shall provide suitable prophylactics for the Contractor's Personnel and shall comply with all the regulations of the local health authorities, including use of appropriate insecticide.

EXAMPLE SUB-CLAUSE

Alcoholic Liquor or Drugs

The Contractor shall not, otherwise than in accordance with the Lawsof the Country, import, sell, give, barter or otherwise dispose of any alcoholic liquor or drugs, or permit or allow importation, sale, gift, barterer disposal by Contractor's Personnel.

EXAMPLE SUB-CLAUSE

Arms and Ammunition

The Contractor shall not give, barter or otherwise dispose of to any person, any arms or ammunition of any kind, or allow Contractor's Personnel to do so.

EXAMPLE SUB-CLAUSE

Festivals and Religious Customs

The Contractor shall respect the Country's recognized festivals, days of rest and religious or other customs.

Clause 7 Plant, Materials and Workmanship

Additional Sub-Clause

If the Contract is being financed by an institution whose rules or policies require a restriction on the use of its funds, a further sub- clause may be added:

EXAMPLE SUB-CLAUSE

All Goods shall have their origin in eligible source countries as defined in (insert name of published guidelines for procurement).

Goods shall be transported by carriers from these eligible source countries, unless exempted by the Engineer in writing on the basis of potential excessive costs or delays. Surety, insurance and banking services shall be provided by insurers and bankers from the eligible source countries.

Clause 8Commencement, Delays and Suspension

Sub-Clause 8.2 Time for Completion

If the Works are to be taken-over in stages, these stages should be defined as Sections, in the Appendix to Tender.

Sub-Clause 8.7 Delay Damages

Under many legal systems, the amount of these pre-defined damages must represent a reasonable pre-estimate of the Engineer's probableloss in the event of delay. If the Accepted Contract Amount is to be quoted as the sum of figures in more than one currency, it may be preferable to define these damages (per day) as the percentage reduction which would be applied to each of these figures. If the Accepted Contract Amount is expressed in the Local Currency, the damages per day may either be defined as a percentage or be definedas a figure in Local Currency: see Sub-Clause 14.15(b).

Additional Sub-Clause

Incentives for early completion may be included in the tender documents (although Sub-Clause 13.2 refers to accelerated completion):

EXAMPLE SUB-CLAUSE

Sections are required to be completed by the dates given in the Appendix to Tender in order that these Sections may be occupiedand used by the Engineer in advance of the completion of the whole of the Works. Details of the work required to be executed to entitle the Contractor to bonus payments and the amount of the bonuses are stated in the Specification.

For the purposes of calculating bonus payments, the dates given in the Appendix to Tender for completion of Sections are fixed. No adjustments of the dates by reason of granting an extension of the Time for Completion will be allowed.

Clause 9Tests on Completion

Sub-Clause 9.1 Contractor's Obligations

The Specification should describe the tests which the Contractor is to carry out before being entitled to a Taking-Over Certificate. If the Works are to be tested and taken-over in stages, the tests requirements may have to take account of the effect of some parts ofthe Works being incomplete.

Clause 10 Engineer's Taking Over

Sub-Clause 10.1 Taking-Over Certificate

If the Works are to be taken-over in stages, these stages should tobe defined as Sections, in the Appendix to Tender. Precise geographical definitions are advisable, and the Appendix shouldinclude a table, so as to define the Time for Completion and delay damages: the table is shown in the example Appendix.

Clause 11 Defects Liability

Sub-Clause 11.10 Unfulfilled Obligations

It may be necessary to review this Sub-Clause for the period of liability imposed by the applicable law.

Clause 12 Measurement and Evaluation

Sub-Clause 12.1 Works to be Measured

If any part of the Permanent Works is to be measured according to records of its construction, details should be specified in the tender documents, including any records for which the Contractor is to be responsible.

Clause 13 Variations and Adjustments

Variations can be initiated by any of three ways:

(a) the Engineer may instruct the variation under Sub-Clause 13.1, without prior agreement as to feasibility or price;

(b) the Contractor may initiate his own proposals under Sub-Clause 13.2, which are intended to benefit both Parties; or

(c) the Engineer may request a proposal under Sub-Clause 13.3,seeking prior agreement so as to minimize dispute.

Sub-Clause 13.8 Adjustments for Changes in Cost

These provisions for adjustments may be required if it would be unreasonable for the Contractor to bear the risk of escalating costs due to inflation. Unless this Sub-Clause is not to apply, the Appendix to Tender should include a table for each of the currencies of payment: the appropriate table is shown in the example Appendix.Particular care should be taken in the calculation of the weightings/coefficients ("a", "b", "c", ..., the total of which must not exceed unity) and in the selection and verification of cost indices. Expert advice may be appropriate.

Clause 14 Contract Price and Payment

Sub-Clause 14.1 The Contract Price

When writing the Particular Conditions, consideration should be given to the amount and timing of payment(s) to the Contractor. A positive cash flow is clearly of benefit to the Contractor, and tenderers will take account of the interim payment procedures when preparing their tenders.

Additional Sub-Clauses may be required to cover any exceptions to the options set out in Sub- Clause 14.1, and any other mattersrelating to payment.

Cost-plus contracts, under which the actual Costs are determined and paid, are unusual and only used when (for reasons of urgency or otherwise) the Engineer is willing to accept the risks involved. Ifthe Contractor is to be paid actual Costs, Clause 12 should bereplaced by provisions describing the method of determining the Costs and Contract Price. As a result, the provisions in the General Conditions which entitle the Contractor to payment of additional Costs will generally be of no effect.

Sub-Clause 14.1(a) would not apply if payment is to be made on a lump sum basis.

Lump sum contracts may be suitable if the tender documents include details which are sufficiently complete for construction and for Variations to be unlikely.

From the information supplied in the tender documents, the Engineer can prepare the design and cost and any other details necessary, and Contractor construct the Works, with having to refer back to the Engineer for clarification or further information.

Further design by the Engineer (under sub-paragraphs (a) to (d) of Sub-Clause 4.1) is not precluded.

However, these Conditions would be inappropriate if significant design input by the Contractor is required. In those cases, ELF's other forms may be more appropriate: see ELF's Conditions of Contract for Plant and Design-Build or Conditions of Contract forTurnkey Projects.

In case of above also Engineer will design and prepare the cost and suitable condition of contract as per suitability for Turnkey or PPP model or Plant and Design Built Contract.

In case few works need Turnkey then Lumpsum Cost is to be given to contractor for particular work.

If work needs Public Private Partnership, then Engineer will Design and find out the cost and basis on this few works should be included in the contract similarly in Plant and Design-Built contract.

For a lump sum contract, the tender documents should include a schedule of payments (see Sub- Clause 14.4), and any drawings required for construction may be specified as being Contractor's Documents. The Specification should describe the procedures under which the Contractor submits these Documents for the Engineer to approve.

EXAMPLE PROVISIONS FOR A LUMP SUM CONTRACT

Delete Clause 12.

Delete the last sentence of Sub-Clause 13.3 and substitute:

Upon instructing or approving a Variation, the Engineer shall proceed in accordance with Sub-Clause 3.5 to agree or determine adjustments to the Contract Price and to the schedule of paymentsunder Sub-Clause 14.4. These adjustments shall include reasonable profit, and shall take account of the Contractor's submissions under Sub-Clause 13.2 if applicable.

Delete sub-paragraph (a) of Sub-Clause 14.1 and substitute:

(a) the Contract Price shall be the lump sum Accepted Contract Amount and be subject to adjustments in accordance with the Contract;

(b) If Sub-Clause 14.1(b) is not to apply, additional Sub-Clause(s) should be added.

EXAMPLE SUB-CLAUSE ON EXEMPTION FROM DUTIES

All Goods imported by the Contractor into the Country shall be exempt from customs and other import duties, if the Engineer's prior written approval is obtained for import. The Engineer shall endorse the necessary exemption documents prepared by the Contractor for presentation in order to clear the Goods throughCustoms, and shall also provide the following exemption documents:

(Describe the necessary documents, which the Contractor will be unable to prepare)

If exemption is not then granted, the customs duties payable and paid shall be reimbursed by the Engineer.

All imported Goods, which are not incorporated in or expended in connection with the Works, shall be exported on completion of the Contract. If not exported, the Goods will be assessed for duties as applicable to the Goods involved in accordance with the Laws ofthe Country.

However, exemption may not available for:

(a) Goods which are similar to those locally produced, unless they are not available in sufficient quantities or are of a different standard to that which is necessary for the Works; and

(b) any element of duty or tax inherent in the price of goods or services procured in the Country, which shall be deemed to be included in the Accepted Contract Amount.

Port dues, quarry dues and, except as set out above, any element of taxor duty inherent in the price of goods or services shall be deemedto be included in the Accepted Contract Amount.

EXAMPLE SUB-CLAUSE ON EXEMPTION FROM TAXES

Expatriate (foreign) personnel shall not be liable for income tax levied in the Country on earnings paid in any foreign currency, or for income tax levied on subsistence, rentals and similar services directly furnished by the Contractor to Contractor's Personnel, or for allowances in lieu. If any

Contractor's Personnel have part of their earnings paid in the Country in a foreign currency, they may export (after the conclusionof their term of service on the Works) any balance remaining of theirearnings paid in foreign currencies.

The Engineer shall seek exemption for the purposes of this Sub-Clause. If it is not granted, the relevant taxes paid shall be reimbursed by the Engineer.

Sub-Clause 14.2Advance Payment

When writing the Particular Conditions, consideration should be given to the benefits of advance payment(s). Unless this Sub-Clauseis not to apply, the total advance payment (and the number of instalments if more than one) must be specified in the Appendix to Tender. Threat of deduction for the repayments should be checked to ensure that repayment is achieved before completion. The typical figures in sub-paragraphs (a) and (b) of the General Conditions Sub- Clause are based on the assumption that the total advance payment is less than 22% of the Accepted Contract Amount.

The acceptable form(s) of guarantee should be included in the tender documents, annexed to the Particular Conditions: an example formis annexed to this document, as Annex E.

Sub-Clause 14.7 Payment

If a different period for payment is to apply, the Sub-Clause may be amended:

EXAMPLE

In sub-paragraph (b) of Sub-Clause 14.7, delete "56" and substitute "42"

If the country/countries of payment need to be specified, details maybe included in a Schedule.

Sub-Clause 14.8 Delayed Payment

If the discount rate of the central bank in the country of the currency of payment is not a reasonable basis for assessing the Contractor's financing costs, a new rate may have to be defined. Alternatively, the actual financing Costs could be paid, taking account of local financing arrangements.

Sub-Clause 14.9 Payment of Retention Money

If part of the Retention Money is to be released and substituted byan appropriate guarantee, an additional Sub-Clause may be added. The acceptable form(s) of guarantee should be included in the tender documents, annexed to the Particular Conditions: an example form is annexed to this document, as Annex F.

EXAMPLE SUB-CLAUSE FOR RELEASE OF RETENTION

When the Retention Money has reached three-fifths (60%) of the limit of Retention Money stated in the Appendix to Tender, the Engineer shall certify and the Engineer shall make payment of half (50%) ofthe limit of Retention Money to the Contractor if he obtains a guarantee, in a form and provided by an entity approved by the Engineer, in amounts and currencies equal to the payment.

The Contractor shall ensure that the guarantee is valid and enforceable until the Contractor has executed and completed the Works and remedied any defects, as specified for the Performance Security in Sub-Clause 4.2, and shall be returned to the Contractor accordingly. This release of retention shall be in lieu of the release ofthe second half of the Retention Money under the second paragraph of Sub-Clause 14.9.

Sub-Clause 14.15 Currencies of Payment

If all payments are to be made in Local Currency, it must be named in the Letter of Tender, and only the first sentence of this Sub-Clause will apply. Alternatively, the Sub-Clause may then be replaced:

EXAMPLE SUB-CLAUSE FOR A SINGLE CURRENCY CONTRACT

The currency of account shall be the Local Currency and all payments made in accordance with the Contract shall be in Local Currency. The Local Currency payments shall be fully convertible, except those for local costs. The percentage attributed to local costs shall be as stated in the Appendix to Tender.

Financing Arrangements

For major contracts in some markets, there may be a need to secure finance from entities such as aid agencies, development banks, export credit agencies, or other international financing institutions.If financing is to be procured from any of these sources, the Particular Conditions may need to incorporate its special requirements. The exact wording will depend on the relevant institution, so reference will need to be made to them to ascertaintheir requirements, and to seek approval of the draft tender documents.

These requirements may include tendering procedures which need tobe adopted in order to render the eventual contract eligible forfinancing, and/or special Sub-Clauses which may need to be incorporated into the Particular Conditions. The following examples indicate some of the topics which the institution's requirements may cover:

(a) prohibition from discrimination against the shipping companies of any one country;

(b) ensuring that the Contract is subject to a widely- accepted neutral law;

(c) prevision for arbitration under recognized international rules and at a neutral location;

(d) giving the Contractor the right to suspend/terminate in the event of default under the financing arrangements;

(e) restricting the right to reject Plant;

(f) spacifying the payments due in the event of termination;

(g) specifying that the Contract does not become effective until certain conditions precedent have been satisfied, including pre-disbursement conditions for the financing arrangements; and

(h) obliging the Employer to make payments from his own resources if, for any reason, the funds under the financing arrangements are insufficient to meet the payments due to the Contractor, whether due to a default under the financing arrangements or otherwise.

In addition, the financing institution or bank may wish the Contract to include references to the financing arrangements, especially if funding from more than one source is to be arrangedto except: example text between lines may be copied finance different elements of supply. It is not unusual for the Particular Conditions to include special provisions identifying differentcategories of Plant and specifying the documents to be presented to the relevant financing institution to obtain payment. If the financing institution's requirements are not met, it may be difficult (or even impossible) to secure suitable financing for the project, and/or the institution may decline to provide finance for part or all of the Contract.

However, where the financing is not tied to the export of goods and services from any particular country but is simply provided by commercial banks' lending to the Engineer, those banks may be concerned to ensure that the Contractor's rights are very restricted.These banks may wish the Contract to exclude any reference to the financing arrangements, and/or to restrict the Contractor's rights under Clause 16.

FORM OF SUB-CLAUSE WHICH A FINANCING INSTITUTION MAY REQUIRE

The Accepted Contract Amount is made up as follows:(breakdown into items and/or into supply/delivery/etc.)

and shall be payable by the Engineer to the Contractor as set out below.

(a)

_____% of the Accepted Contract Amount shall be payable by a direct payment from the Engineer to the Contractor within 28 days of receipt by the Engineer of the following documents:

(a) commercial invoice addressed to the Engineer specifying the amount of the payment now due,

(ii) advance payment security guarantee issued by_______________________________________Bank in theform annexed,

(iii) performance security guarantee issued by _Banking the formannexed, and

(iv) Interim Payment Certificate confirming the payment due and specifying the amount

(b)

____% of the contract price for the supply of Plant shall be payable as follows:

(i) _____ % of the estimated contract value of the Plant supplied, bydirect payment from the Employer to the Contractor on shipment

of each item, against the following documents:

(original) commercial invoice,

(original) shipping documents,

(original) certificate of origin,(original) insurance certificate, and

(original) Interim Payment Certificate confirming the payment dueand specifying the amount.

(ii) _____ % of the estimated contract value of the Plant supplied, by disbursement from the Loan Agreement to the Contractor on shipment of each item, on presentation of a Qualifying Certificate in the form annexed and copies of the documents listed in sub- paragraph (b)(I) above.

(c) the balance of the Contract Price shall be payable as follows:

(i) _____ % of the estimated contract value of the ervicesrendered, by direct payment from the Engineer to the contractor on execution of the relevant service, against the following documents:

(original) commercial invoice, and

(original) Interim Payment Certificate confirming the payment due and specifying the amount.

(ii) _____ % of the estimated contract value of the services rendered, by disbursement from the Loan Agreement to the Contractor, on presentation of a Qualifying Certificate in the form annexed and copies of the documents listed in sub-paragraph (c)(I) above.

(d)

he direct payments by the Engineer specified in sub-paragraph

(b) shall be made by an irrevocable letter of credit established by the Employer in favor of the Contractor and confirmed by a bank acceptable to the Contractor.

The above arrangements (involving financing institution(s), Engineer and Contractor) may be initiated by the Engineer; or bythe Contractor, before submitting the Tender. Alternatively, the Contractor may be prepared to initiate financing arrangements and retain responsibility for them, although he would probably be unable or unwilling to provide finance from his own resources. Hisfinancing bank's requirements would then affect his attitude in contract negotiations. They might well require the Engineer to make interim payments, although a large proportion of the Contract Price might be withheld until the Works are complete.

This payment arrangement can be achieved either by a high Percentage of Retention; or by a suitably completed schedule of payments (see Sub-Clause 14.4), with the Instructions to Tenderers specifying the criteria with which the Tenderer should comply. Since the Contractor would then have to arrange his own financing to cover the shortfall between the payments and his outgoings, he (and his financing bank) would probably require some form of security, guaranteeing payment when due.

It may be appropriate for the Engineer, when preparing the tender documents, to anticipate the latter requirement by undertaking to provide a guarantee for the element of payment which the Contractor is to receive when the Works are complete. The acceptable form(s) of guarantee should be included in the tender documents, annexed to the Particular Conditions: an example form is annexed to this document, as Annex G. The following

Sub-Clause may be added.

EXAMPLE PROVISIONS FOR CONTRACT OR FINANCE

The Engineer shall obtain (at his cost) a payment guarantee in the amount and currencies, and provided by an entity, as stated in the Appendix to Tender. The Engineer shall deliver the guarantee to the Contractor within 28 days after both Parties have entered into the Contract Agreement. The guarantee shall be in the formannexed to these Particular Conditions, or in another form acceptable to the Contractor. Unless and until the Contractor receives the guarantee, the Engineer shall not give the notice underSub-Clause 8.1.

The guarantee shall be returned to the Engineer at the earliest of the following dates:

(a) when the Contractor has been paid the Accepted Contract Amount;

(b) when obligations under the guarantee expire or have been discharged; or

(c) when the Employer has performed all obligations under the Contract.

Clause 15 Termination by Engineer

Sub-Clause 15.2 Termination by Engineer

Before inviting tenders, the Engineer should verify that the wording of this Sub-Clause, and each anticipated ground for termination, is consistent with the law governing the Contract.

Sub-Clause 15.5 Engineer's Entitlement to Termination

Unless inconsistent with the requirements of the Engineer and/or financing institutions, a further sentence may be added.

EXAMPLE Insert at the end of Sub-Clause 15.5:

The Engineer shall also pay to the Contractor the amount of any other loss or damage resulting from this termination.

Clause 16 Suspension and Termination by Contractor

Sub-Clause 16.2 Termination by Contractor

Before inviting tenders, the Engineer should verify that the wording of this Sub-Clause is consistent with the law governing the Contract. The Contractor should verify that each anticipated ground for termination is consistent with such law.

Clause 17 Risk and Responsibility

Sub-Clause 17.6 Limitation of Liability

EXAMPLE

In Sub-Clause 17.6, the sum referred to in the penultimatesentence shall be __ Additional Sub-Clause Use of Engineer's Accommodation/Facilities

If the Contractor is to occupy the Employer's facilities temporarily,an additional sub-clause may bead:

EXAMPLE SUB-CLAUSE

The Contractor shall take full responsibility for the care of theitems detailed below, from the respective dates of use or occupation by the Contractor, up to the respective dates of hand-over or cessation of occupation (where hand-over or cessation of occupation may take place after the date stated in the Taking-Over Certificate for the Works):

(insert details)

If any loss or damage happens to any of the above items while the Contractor is responsible for their care, arising from any cause whatsoever other than those for which the Engineer is liable, the Contractor shall, at his own cost, rectify the loss or damage to the satisfaction of the

Engineer.

Clause 18 Insurance

The wording in the General Conditions describes the insurances which are to be arranged by the "insuring Party", who is to be the Contractor unless otherwise stated in the Particular Conditions. Insurances so provided by the Contractor are to be consistent withthe general terms agreed with the Engineer. The Instructions to Tenderers may therefore require tenderers to provide details of the proposed terms.

If the Engineer is to arrange any of the insurances under this Clause,the tender documents should include details as an annex to theParticular Conditions (so that tenderers can estimate what otherinsurances they wish to have for their own protection), including the conditions, limits, exceptions and deductibles; preferably in the form of a copy of each policy. The Engineer may find it difficult to affect the insurances described in the third paragraph of Sub-Clause

18.2 (for Contractor's Equipment, which includes Subcontractor's equipment), because the Engineer may not know the amount or value of these items of equipment. The following sentence may beincluded in the Particular Conditions:

EXAMPLE Delete the final paragraph ofSub-Clause 18.2 and substitute:

However, the insurances described in the first two paragraphs of Sub-Clause 18.2 shall be affected and maintained by the Engineeras insuring Party, and not by the Contractor.

Clause 19 Force Majeure

Before inviting tenders, the Engineer should verify that the wording ofthis Clause is compatible with the law governing the Contract.

Clause 20 Claims, Disputes and Arbitration

Sub-Clause 20.2 Appointment of the Dispute ResolutionCommittee

Unless the Engineer (although appointed by the Employer) is to make the pre-arbitral decisions under this Clause 20, in accordance with the alternative option described below, the Contract should include the provisions under Clause 20 which, whilst not discouraging the Parties from reaching agreement on disputes as the works proceed, allow them to refer contentious matters to an impartial dispute resolution committee ("DRC").

The adjudication procedure depends for its success on, amongst other things, the Parties' confidence in the agreed individual(s) who will serve on the DRC. Therefore, it is essential that candidates for this position are not imposed by either Party on the other Party; and that, if the individual is selected under Sub-Clause 20.3, the selection is made by a wholly impartial entity. ELF is prepared to perform this role, if this authority has been delegated in accordance with the example wording in the Appendix to Tender.

It is preferable, but not essential, for the individual(s) to be agreed before the Letter of Acceptance is issued, and for the DRC to visit the Site on a regular basis. Under the example text in the Appendixto Tender, the Parties may either so agree before the Letter of Acceptance is issued or agree the appointment within the specified period thereafter. Alternatively, the Parties may prefer to defer theappointment until a dispute has arisen, in which case Sub-Clause

20.2 plus the Appendix - General Conditions of Dispute Resolution Committee with its Annex (Procedural Rules) and the Dispute Adjudication Agreement should be amended to comply with the wording

except: example text between lines may be copied 19contained in the corresponding sections.

Sub-Clause 20.2 provides for two alternative arrangements for the DAB:

(a) one person, who acts as the sole member of the DRC, having entered into a tripartite agreement with both Parties; or

(b)

DRC of six persons, each of whom has entered into a tripartite agreement with both Parties.

The form of this tripartite agreement could be one of the two alternatives shown at the end of this publication, as appropriate to the arrangement adopted. Both of these forms incorporate (by reference) the General Conditions of Dispute Adjudication Agreement, which are included as the Appendix to the General Conditions because they are also referred to in Sub-Clause 20.2. Under either of these alternative forms of Dispute Adjudication Agreement, each individual person is referred to as a member.

At an early stage, consideration should be given as to whether a six-person DRC is preferable for a particular project, taking account of its size, duration and the fields of expertise which will be involved. For some projects, it may be considered appropriate to appoint a six- person DRC for each major field of expertise relevant to the Works; however, this may give rise to problems if, when a dispute arises, the Parties cannot agree which field is applicable and, there- fore, to whom the dispute should be referred.

For a six-person DRC to be mutually agreed, the Engineer (or the tenderer) could propose the names and curriculum vitae of suitablepersons, for the tenderer (or the Engineer) to accept. It may be advisable to propose alternates in case some subsequently decline the appointment, assuming that they have not previously indicated their willingness to accept. Each Party may be reluctant to choose names from a list of people who have already been contacted by the other Party.

For a six-person DRC, the Engineer and the tenderer may each propose two members, similar to the above procedure, for the tendererand the Engineer respectively to accept. For the chairman, the Engineer (or the tenderer) could similarly propose suitable persons for the tenderer (or the Engineer) to accept. It may be appropriate for the chairman's retainer fee to be more than that of the other two members,reflecting the additional administrative tasks which a chairman will have to perform.

The appointment of the DRC may be facilitated, especially if the members are not to be appointed at the commencement of the Contract, by including an agreed list of potential members in the Contract: in a Schedule.

Alternatively, the Engineer may make these pre-arbitral decisions. This alternative, which has been the Engineer's traditional role in common law countries, may be appropriate if the Engineer is an independent professional consulting engineer with the experience and resources required for the administration of all aspects of the contract. The Employer should recognize that, although the Engineer generally acts for the Employer as specified in Sub-Clause 3.1(a), the Engineer will make these pre-arbitral decisions fully and the Employer must not prejudice this impartiality. If this alternativeis considered appropriate, the Sub-Clause may be varied:

EXAMPLE SUB-CLAUSE FOR PRE-ARBITRAL DECISIONS BYTHE

ENGINEER

Delete Sub-Clauses 20.2 and 20.3.

Delete the second paragraph of Sub-Clause 20.4 and substitute:

The Engineer shall act as the DRC in accordance with this Sub-Clause20.4, acting fairly, impartially and at the cost of the Employer.

In the event that the Employer intends to replace the Engineer, the Employer's notice under Sub-Clause 3.4 shall include detailed proposals for the appointment of a replacement DRC.

Sub-Clause 20.5 Amicable Settlement

The provisions of this Sub-Clause are intended to encourage the parties to settle a dispute amicably, without the need for arbitration: for example, by direct negotiation, conciliation, mediation, or other forms of alternative dispute resolution. Amicable settlement procedures often depend, for their success, on confidentiality and on both Parties' acceptance of the procedure. Therefore, neither Party should seek to impose the procedure on the other Party.

Sub-Clause 20.6 Arbitration

The Contract should include provisions for the resolution by international arbitration of any disputes which are not resolved amicably. In international contracts, international commercial arbitration has numerous advantages over litigation in national courts, and may be more acceptable to the Parties.

Careful consideration should be given to ensuring that the international arbitration rules chosen are compatible with the provisions of Clause 20 and with the other elements to be set out in the Appendix to Tender. The Rules of Arbitration of the International Chamber of Commerce (the "ICC", which is based at 38 Cours Albert 1er, 75008 Paris, France) are frequently included in international contracts. In the absence of specific stipulations as to the number of arbitrators and the place of arbitration, the International Court of Arbitration of the ICC will decide on the number of arbitrators (typically three in any substantial construction dispute) and on the place of arbitration.

If the UNCITRAL (or other non-ICC) arbitration rules are preferred,it may be necessary to designate, in the Appendix to Tender, an institution to appoint the arbitrators or to administer the arbitration, unless the institution is named (and their role specified)in the arbitration rules. It may also be necessary to ensure, beforeso designating an institution in the Appendix to Tender, that it is prepared to appoint or administer.

For major projects tendered internationally, it is desirable that the place of arbitration be situated in a country other than that of the Employer or Contractor. This country should have a modern and liberal arbitration law and should have ratified a bilateral or multilateral convention (such as the 1958New York Convention on the Recognition and Enforcement of Foreign Arbitral Awards), or both,that would facilitate the enforcement of an arbitral award in the states of the Parties.

It may be considered desirable in some cases for other Parties to be joined into any arbitration between the Parties, thereby creating a multi-party arbitration. While this may be feasible, multi- party arbitration clauses require skillful drafting, and usually need to be prepared on a case-by-case basis.

21. MAINTENANCE

Details of Maintenance including daily activity, monthly activityand all maintenance points to be considered.

Payments of maintenance should base on cost plus basis. 21.1-

Maintenance obligations of the Contractor 22. Supervision and Monitoring during Maintenance

22.1 Inspection by the Contractor

Annexes FORMS OF SECURITIES

Acceptable form(s) of security should be included in the tender documents: for Annex A and/or B, in the Instructions to Tenderers; and for Annexes C to G, annexed to the Particular Conditions. The following example forms, which (except for Annex A) incorporate Uniform Rules published by the International Chamber of Commerce (the "ICC", which is based at 38 Cours Albert 1er, 75008 Paris, France), may have to be amended to comply with the applicable law. Although the ICC publishes guides to these Uniform Rules, legal advice should be taken before the securities are written. Note that the guaranteed amounts should be quoted inall the currencies, as specified in the Contract, in which the guarantor pays the beneficiary.

Annex An EXAMPLE FORM OF PARENT COMPANY GUARANTEE

[*See page 3, and the comments on Sub-Clause 1.14*]

Brief description of Contract_____________________________Name and address of Engineer__________________________(together with successors and assigns).

We have been informed that (hereinafter called the "Contractor") is submitting an offer for such Contract in response to your invitation, and that the conditions of your invitation require his offer to be supported by a parent company guarantee.

In consideration of you, the Engineer, awarding the Contract to the Contractor, we (*name of parent company*) i r r e v o c a b l y and unconditionally guarantee to you, as a primary obligation, the due performance of all the Contractor's obligations and liabilities under the Contract, including the Contractor's compliance with all its terms and conditions according to their true intent and meaning.

If the Contractor fails to so perform his obligations and liabilities and comply with the Contract, we will indemnify the Engineer against and from all damages, losses and expenses (including legal fees and expenses) which arise from any such failure for which the Contractor is liable to the Employer under the Contract.

This guarantee shall come into full force and effect when the Contract comes into full force and effect. If the Contract does not come into full force and effect within a year of the date of this guarantee, or if you demonstrate that you do not intend to enterinto the Contract with the Contractor, this guarantee shall be void and ineffective. This guarantee shall continue in full force and effect until all the Contractor's obligations and liabilities under the Contract have been discharged, when this guarantee shall expire andshall be returned to us, and our liability hereunder shall be discharged absolutely.

This guarantee shall apply and be supplemental to the Contract as amended or varied by the Engineer and the Contractor from time to time. We hereby authorize them to agree any such amendmentor variation, the due performance of which and compliance with which by the Contractor are likewise guaranteed hereunder. Our obligations and liabilities under this guarantee shall not be discharged by any allowance of time or other indulgence whatsoever by the Engineer to the Contractor, or by any variationor suspension of the works to be executed under the Contract, or by any amendments to the Contract or to the constitution of the Contractor or the Engineer, or by any other matters, whether withor without our knowledge or consent.

This guarantee shall be governed by the law of the same country (or other jurisdiction) as that which governs the Contract and any dispute under this guarantee shall be finally settled under the Rules of Arbitration of the International Chamber of Commerce by one or more arbitrators appointed in accordance with such Rules. We confirm that the benefit of this guarantee may be assigned subject only to the provisions for assignment of the Contract.

Date_______________ Signature(s)____________________________

23

Annex B EXAMPLE FORM OF TENDERSECURITY

[*See page 4*]

Brief description of Contract ____________

Name and address of Beneficiary _

(whom thetender documents define as the Employer).

We have been informed that _

____

(hereinafter called the "Principal") is submitting an offer for such Contract in response to your invitation, and that the conditions ofyour invitation (the "conditions of invitation", which are set out in a document entitled Instructions to Tenderers) require his offer to besupported by a tender security.

At the request of the principal, we (*name of bank*) ____Hereby irrevocably undertake to pay you, the Beneficiary/Employer, any sum or sums not exceeding in total the amount of____(say:______ ____)upon receipt by us of your demand in writing and your written statement (in the demand) stating that:

a. the principal has, without your agreement, withdrawn his offerafter the latest time specified for its submission and before theexpiry of its period of validity, or

b. the principal has refused to accept the correction of errors inhis offer in accordance with such conditions of invitation, or

c. you awarded the Contract to the Principal and he has failed to comply with sub-clause 1.6of the conditions of the Contract, or

d. you awarded the Contract to the Principal and he has failed to comply with sub-clause 4.2of the conditions of the Contract.

Any demand for payment must contain your signature(s) which must be authenticated by your bankers or by a notary public. The authenticated demand and statement must be received by us at thisoffice on or before (*the date 35 days after the expiry of the validity ofthe Letter of Tender*)

____________________________________, when this guarantee shall expire and shall be returned to us.

This guarantee is subject to the Uniform Rules for Demand Guarantees, published by the International Chamber of Commerce, except as stated above.

Date__________________________________Signature(s) _

____

Annex C: EXAMPLE FORM OFPERFORMANCE SECURITY - DEMAND GUARANTEE

[*See comments on Sub-Clause 4.2*]

Brief description of Contract ____________

Name and address of Beneficiary _

(whom the Contract defines as the Engineer).

We have been informed that

(hereinafter called the "Principal") is your contractor under such Contract, which requires him to obtain a performance security.

At the request of the Principal, we (*name of bank*) ________________________________hereby irrevocably undertake to pay you, the Beneficiary/Engineer, any sum or sums not exceeding in total the amount of__________(the "guaranteed amount", say:

upon receipt by us of your demand in writing and your written statement stating:

(a) that the principal is in breach of his obligation(s) under the Contract,and

(b) the respect in which the principal is in breach.

[Following the receipt by us of an authenticated copy of the taking- over certificate for the whole of the works under clause 10 of the conditions of the Contract, such guaranteed amount shall bereduced by % and we shall promptly notify you that we have received such certificate and have reduced the guaranteed amountaccordingly.] (1)

Any demand for payment must contain your [minister's/directors'] (1) signature(s) which must be authenticated by your bankers or bya notary public. The authenticated demand and statement must be received by us at this office on or before (*the date 70 days after the expected expiry of the Defects Notification Period for the Works*)

(the "expiry date"), when this guarantee shall expire and shall be returned to us.

We have been informed that the Beneficiary may require the principalto extend this guarantee if the performance certificate under the Contract has not been issued by the date 28 days prior to such expiry date. We undertake to pay you such guaranteed amount upon receipt by us, within such period of 28 days, of your demand in writing and your written statement that the performancecertificate has not been issued, for reasons attributable to the principal, and that this guarantee has not been extended.

This guarantee shall be governed by the laws ofand shall be subject to the Uniform Rules for Demand Guarantees, published by the International Chamber of Commerce, except as stated above.

Date________________________________Signature(s) _

____

When writing the tender documents, the writer should ascertain whetherto include the optional text, shown in parentheses [ ]25

Annex DEXAMPLE FORM OFPERFORMANCE SECURITY - SURETY BOND

[*See comments on Sub-Clause 4.2*]

Brief description of Contract _

Name and address of Beneficiary _

____

______________________________(together with successors and assigns, all as defined in the Contract as the Employer).

By this Bond, (*name and address of contractor*)

____

(who is the contractor under such Contract) as Principal and (*name and address of guarantor*)

as ______________________________ Guarantor are irrevocably held and firmly bound to the Beneficiary in the total amount of

____

____

(the"Bond Amount", say:

______________________________) For the due performance of all such Principal's obligations and liabilities under the Contract. [Such Bond Amount shall be reduced by % upon the issue of the taking-over certificate for the whole of the works under clause 10 ofthe conditions of the Contract.] (1)

This Bond shall become effective on the Commencement Date defined in the Contract.

Upon Default by the Principal to perform any Contractual Obligation,or upon the occurrence of any of the events and circumstances listed in sub-clause 15.2 of the conditions of the Contract, the Guarantor shall satisfy and discharge the damages sustained by the Beneficiary due tosuch Default, event or circumstances. (2) However, the total liability of the Guarantor shall not exceed the Bond Amount.

The obligations and liabilities of the Guarantor shall not be discharged by any allowance of time or other indulgence whatsoever by the Beneficiary to the Principal, or by any variation or suspensionof the works to be executed under the Contract, or by any amendments to the Contract or to the constitution of the principal or the Beneficiary, or by any other matters, whether with or without the knowledge or consent of the Guarantor.

Any claim under this Bond must be received by the Guarantor on or before (*the date six months after the expected expiry of the Defects Notification Period for the Works*)

(the "Expiry Date"), when this Bond shall expire and shall be returned to the Guarantor.

The benefit of this Bond may be assigned subject to the provisions for assignment of the Contract, and subject to the receipt by the Guarantor of evidence of full compliance with such provisions.

This Bond shall be governed by the law of the same country (or another jurisdiction) as that which governs the Contract. This Bond incorporates and shall be subject to the Uniform Rules for Contract Bonds, published by the International Chamber of Commerce, and words used in this Bond shall bear the meanings set out in such Rules.

Wherefore this Bond has been issued by the Principal and the Guarantor on (*date*) Signature(s)

for and on behalf of the Principal

Signature(s) for and on behalf of the Guarantor _

When writing the tender documents, the writer should ascertain whetherto include the optional text, shown in parentheses []

Insert: [and shall not be entitled to perform the principal's obligations under the Contract.]

Or: [or at the option of the Guarantor (to be exercised in writing within 42 days of receiving the claim specifying such Default) perform the principal's obligations under the Contract.]

Annex E EXAMPLE FORM OF ADVANCE PAYMENT GUARANTEE

[*See comments on Sub-Clause 14.2*]

Brief description of Contract____________

Name and address of Beneficiary _

(Whom the Contract defines as the Engineer).

We have been informed that_

____

(hereinafter called the "Principal") is your contractor under such Contract and wishes to receive an advance payment, for which the Contract requires him to obtain a guarantee.

At the request of the Principal, we (*name of bank*) ____hereby irrevocably undertake to pay you, the Beneficiary/Engineer, any sum or sums not excee____ ding in total the amount of________________________ (the "guaranteed amount", say:) upon receipt by us of your demand in writing and your written statement stating:

a. that the principal has failed to repay the advance payment in accordance with the conditions of the Contract, and

b. the amount which the principal has failed to repay.This guarantee shall become effective upon receipt [of the firstinstalment] of the advance payment by the principal. Suchguaranteed amount shall be reduced by the amounts of the advance payment repaid to you, as evidenced by your notices issued under sub-clause 14.6 of the conditions of the Contract. Following receipt (from the principal) of a copy of each purported notice, we shall promptly notify you of the revised guaranteed amount accordingly.

Any demand for payment must contain your signature(s) which must be authenticated by your bankers or by a notary public. The authenticated demand and statement must be received by us at thisoffice on or before (*the date 70 days after the expected expiry of the Time for Completion*)

______________________________________ (the "expiry date"), when this guarantee shall expire and shall be returned to us.

We have been informed that the Beneficiary may require the principalto extend this guarantee if the advance payment has not been repaid by the date 28 days prior to such expiry date. We undertake to pay you such guaranteed amount upon receipt by us, within such period of 28 days, of your demand in writing and your written statement that the advance payment has not been repaid and that this guarantee has not been extended.

This guarantee shall be governed by thelawsof and shall be subject to the Uniform Rules for Demand Guarantees, published by the International Chamber of Commerce, except as stated above.

Date______________________________________Signature(s) _

____

AnnexF EXAMPLE FORM OF RETENTION MONEY GUARANTEE

[*See comments on Sub-Clause 14.9*]

Brief description of Contract____________

Name and address of Beneficiary_

(whom the Contract defines as the Engineer).

We have been informed that_

(hereinafter called the "Principal") is your contractor under such Contract and wishes to receive early payment of [part of] the retention money, for which the Contract requires him to obtain a guarantee.

At the request of the Principal, we (*name of bank*)

____hereby irrevocably undertake to pay you, the Beneficiary/Engineer, any sum or sums not exceding in total the amount of____________________

____

(the "guaranteed amount", say:)upon receipt by us of your demand in writing and your written statement stating:

(a) that the principal has failed to carry out his obligation(s) to rectify certain defect(s) for which he is responsible under the Contract, and

(b) the nature of such defect(s).

At any time, our liability under this guarantee shall not exceed the total amount of retention money released to the principal by you, as evidenced by your notices issued under sub-clause 14.6 of the conditions of the Contract with a copy being passed to us.

Any demand for payment must contain your signature(s) whichmust be authenticated by your bankers or by a notary public. The authenticated demand and statement must be received by us at this office on or before (*the* *date 70 days after the expected expiry of the Defects Notification Period for the Works)*_(the "expiry date"),when this guarantee shall expire and shall be returned to us.

We have been informed that the Beneficiary may require the principalto extend this guarantee if the performance certificate under the Contract has not been issued by the date 28 days prior to such expiry date. We undertake to pay you such guaranteed amount upon receipt by us, within such period of 28 days, of your demand in writing and your written statement that the performancecertificate has not been issued, for reasons attributable to the principal, and that this guarantee has not been extended.

This guarantee shall be governed by the laws of and shall be subject to the Uniform Rules for Demand Guarantees, published by the International Chamber of Commerce, except as stated above.

Date__________________________________Signature(s) _

____

Annex GEXAMPLE FORM OFPAYMENT GUARANTEE BYENGINEER

[*See page 18: Contractor Finance*]

Brief description of Contract__________

Name and address of Beneficiary_

(whomthe Contract defines as the Contractor).

We have been informed that_

(whom the Contract defines as the Engineer and who is hereinafter called the "Principal") is required to obtain a bank guarantee.

At the request of the principal, we (*name of bank*)_

____

____hereby irrevocably undertake to pay you, the Beneficiary/Contractor, any sum or sums not exceeding in total the amount of (say:)

upon receipt by us of your demand inwriting and your written statement stating:

(a) that, in respect of a payment due under the Contract, the Principal has failed to make payment in full by the date fourteen days after the expiry of the period specified in the Contract as that within which such payment should have been made, and

(b) the amount(s) which the principal has failed to pay.

Any demand for payment must be accompanied by a copy of [list of *documents evidencing entitlement to payment*]_______, in respectof which the principal has failed to make payment in full.

Any demand for payment must contain your signature(s) whichmust be authenticated by your bankers or by a notary public. The authenticated demand and statement must be received by us at this office on or before (*the date six months after the expected expiry of the Defects Notification Period for the Works*)__when thisguarantee shall expire and shall be returned to us.

This guarantee shall be governed by the laws of ------------------

and shall be subject to the Uniform Rules for Demand Guarantees, published by the International Chamber of Commerce, except as stated above.

Date________________________________Signature(s) _

____

GENERAL CONDITIONS

GUIDANCE FOR THE PREPARATION OF PARTICULAR CONDITIONS

FORMS OF LETTER OF TENDER, CONTRACT AGREEMENT AND DISPUTE ADJUDICATION AGREEMENT

Conditions of Contract for CONSTRUCTION FOR INFRASTRUCTURE DEVELOPMENT & MAINTENANCEWORK

Forms of Letter of Tender, Contract Agreement and Dispute Adjudication Agreement

Letter Of Tender

NAME OF CONTRACT:

TO:

We have examined the Conditions of Contract, Specification, Drawings, Bill of Quantities, the other Schedules, the attachedAppendix and Addenda Nos

for the execution of the above-named Works. We offer to execute and complete the Works and remedy any defects therein in conformity with this Tender which includes all these documents, for the sum of (in currencies of payment) ______

or such other sum as may be determined in accordance with the Conditions of Contract.

We accept your suggestions for the appointment of the DAB, as set out in Schedule _

[*We have completed the Schedule by adding our suggestions for the other Member of the DAB, but these suggestions are not conditions of this offer].* *

We agree to abide by this Tender until and it shall remain binding upon us and maybe accepted at any time before that date. We acknowledge that the Appendix forms part of this Letter of Tender.

If this offer is accepted, we will provide the specified Performance Security, commence the Workses soon as is reasonably practicable after the Commencement Date, and complete the Works in accordance with the above-named documents within the Time for Completion.

Unless and until a formal Agreement is prepared and executed this Letter of Tender, together with your written acceptance thereof, shall constitute a binding contract between us.

We understand that you are not bound to accept the lowest or any tender you may receive.

Signature____in

the capacity of_________________________________duly

authorized to sign tenders for and on behalf of____________Address:_

____

Date:_______

* If the Tenderer does not accept, this paragraph may be deleted and replaced by:

We do not accept your suggestions for the appointment of the DAB. We have included our suggestions in the Schedule, but these suggestions are not conditions of this offer. If these suggestions arenot acceptable to you, we propose that the DAB be jointly appointed in accordance with Sub-Clause 20.2 of the Conditions of Contract.

APPENDIX TO TENDER

[Note: with the exception of the items for which the Engineer's requirements have been inserted, the following information must be completed before the Tender is submitted]

Item Sub-Clause

Data

Item	Sub-Clause	Data
Employer's name	1.1.2.2 & 1.3	

and address		______
Contractor's name and address	1.1.2.3 & 1.3	______ ______
Engineer's Name and address	1.1.2.4 & 1.3	______ ______ ______
Time for Completion	1.1.3.3.	______

of the Works	. . . .	
Defects Notification n Period	1.1.3.7.	________________
days 365 days Electronic transmissionn systems	1.3	________________
Governing Law	1.4	________________
Ruling language	1.4	________________
Language for communicate	1.4	________________

tons e ……….		
Time for access to the Site ……… …….	2.1…………………. days after Commencement Date	
Amount of Performance	4.2 …………………._________% of the Accepted Contract	

e Security… …..	Amount, in the currencies and proportions in whichthe Contract Price is payable	
Normal working hours ……… ………… …	6.5	________________
Delay damages for the Works ……… ……	8.7 & 14.15(b)	___ % of the final Contract Price per day, in thecurrencies and proportions in which the Contract Price is payable
Maximum amount of delay damages … …..	8.7… …..	________________ `

if there are Provisional Sums: Percentage for adjustment of Provisional Sums	13.5(b).	____% Initial of Signatory of Tender------------------- -----------
If Sub-Clause 13.8 applies: Adjustments for Changes in Cost; Table(s)	13.8	

of adjustment data…….		

8.7 & 14.15(b)__________ % of the final Contract Price per day, in the currencies and proportions in which the Contract Price ispayable

8.7 .... % of the final Contract Price

13.5(b) . . ..______________________________________%

Initial of Signatory of Tender------------------------------

If Sub-Clause 13.8 applies: Adjustments for Changes in Cost; Table(s)of adjustment data . . . .

13.8 . . . .

____

Coefficient; Country of origin; Source of index; Value on stated date(s)*

scope of index currency of index Title/definition Value

Date

a= 0.10 Fixed ______________________________

______ ______ ______ ______

b= Labor______________________________

c= ______________________________

d= ______________________________

e= ______________________________

* These values and dates confirm the definition of each index, butdo not define Base Date indices

Total advance payment……..14.2 . . . .

Number and timing of instalments . . . .14.2 . . . .

Currencies and proportions . . . . .14.2. . . . . . .for

payments each

month/[*YEAR*] in

(*currency*)

Start repayment of advance payment . 14.2(a)____% of the

Accepted Contract Amount

____

Repayment amortization of advance 14.2(b) %

% in ____________

payment.. . . . .

%

Percentage of retention . . . . 14.3 when payments are

%of the

Accepted Contract Limit of Retention Money. . . . .

14.3 .... Amount less Provisional Sums.

If Sub-Clause 14.5 applies:

%

Plant and Materials for 14.5(b) . . . . .

payment when shipped

%

enroute to the Site.

Plant and Materials 14.5(c) . . . .

for payment when deliveredto the Site . . . .

____________% of the Accepted Contract Amount

________________________________[list]

Minimum amount of Interim Payment [list] Certificates. 14.6

________________________________[list]

If payments are only to be made in a currency/currency named on the first page of the Letter of Tender:

Currency/currencies

of payment . . ..   14.15   ______[list]

Initials of signatory of Tender----------------------

If some payments are to be made in a currency/currency not named on the first page of the Letter of Tender:

Currencies of payment 14.15

....[list]

Currency Unit Percentage payable in the Rate of Currency

exchange:

number of

Local per unit of Foreign

Local's,______________[*name*]___________________ 1.000

Foreign:_________[*name*] ______

_________[*name*]___

Periods for submission of insurance:

(a) evidence of insurance 18.1

.... % of the

..............................Accepted Contract Amount

(b) relevant policies . . . . .18.1 . . . .

---

as named in the Letters of Tender

Maximum amount of

18.2(d) . . ..

deductibles for insurance
of the Employer's risks . . ..
Minimum amount of
third party insurance . . . . .18.3 . . . .

Date by which the DAB shall beAppointed

20.2 . . . .

The DAB shall be20.2 . . . .

---

28 days after the
Commencement Date
Appointment (if not agreed)
to be made by . . . . 20.3. . . . . *Either:*
_____________ One sole Member/adjudicator

If there are Sections:
Definition of Sections:

Description	Time for Completion	Delay Damages
(Sub-Clause 1.1.5.6)	(Sub-Clause 1.1.3.3)	(Sub-Clause 8.7)

[*In the above Appendix, the text shown in italics is intended to assist the drafter of a particular contract by providing guidance on which provisions are relevant to the particular contract. This italicized text should not be included in the tender documents, as it will generally appear inappropriate to tenderers.*]

Initials of Signatory of Tender----------------------------

Contract Agreement

This Agreement made the ______________________ day of

____

19 ____

Between ______________________________ of

____

______________________________ (hereinafter called "theEngineer") of the open art ______________, and ________ of

____

Whereas the Engineer desires that the Works known as should be executed by the Contractor, and has accepted a Tender by the Contractor for the execution and completion of these Works and the remedying of any defects therein,

The Engineer and the Contractor agree as follows:

1. In this Agreement words and expressions shall have the same meanings as are respectively assigned to them in the Conditions of Contract hereinafter referred to.
2. The following documents shall be deemed to form and be read and construed as part of this Agreement:

(a) The Letter of Acceptance dated _

____

(b) The Letter of Tender dated _

____

(c) The Addenda no's _____________

(d) The Conditions of Contract

(e) The Specification

(f) The Drawings, and

(g) The completed Schedules.

3. In consideration of the payments to be made by the Engineer to the Contractor as hereinafter mentioned, the Contractor hereby covenants with the Engineer to execute and complete the Works and remedy any defects therein, in conformity with the provisionsof the Contract.

4. The Engineer hereby covenants to pay the Contractor, in consideration of the execution and completion of the Works and the remedying of defects therein, the Contract Price at the times and in the manner prescribed by the Contract.

In Witness whereof the parties hereto have caused this Agreementto be executed the day and

SIGNED by:

____

for and on behalf of the Engineer in the presence of

Witness...

Name...

Address...

Date..

SIGNED by: ______________________________

for and on behalf of the Contractor in the presence of

Witness...

Name...

Address...

Date..

DISPUTE ADJUDICATION AGREEMENT

[for a one-person DAB]

Name and details of Contract ______________________

Name and address of Engineer ______________________

Name and address of Contractor ______________________

Name and address of Member ______________________

Name and address of Engineer ______________________

Name and address of Contractor ______________________

Name and address of Member ______________________

Whereas the Engineer and the Contractor have entered into the Contract and desire jointly to appoint the Member to act as sole adjudicator who is also called the "DAB".

The Engineer, Contractor and Member jointly agree as follows:

1. The conditions of this Dispute Adjudication Agreement comprise the "General Conditions of Dispute Adjudication Agreement", which is appended to the General Conditions of the "Conditions of Contract for Construction" First Edition 2023 published by the Engineer Live Foundation, and the following provisions. In these provisions, which include amendments and additions to the General Conditions of Dispute Adjudication Agreement, words and expressions shall have the same meanings as are assigned to them in the General Conditions of Dispute Adjudication Agreement.
2. [*Details of amendments to the General Conditions of Dispute Adjudication Agreement, if any. For example:*

In the procedural rules annexed to the General Conditions of Dispute

Adjudication Agreement, Rule _ is deleted and replaced by: " ..."]

3. In accordance with Clause 6 of the General Conditions of Dispute Adjudication Agreement, the Member shall be paid as follows:

A retainer fee of______________________________per calendar month,plus a daily fee of______________________per day.

4. In consideration of these fees and other payments to be made by the Engineer and the Contractor in accordance with Clause 6 of the General Conditions of Dispute Adjudication Agreement, the Member undertakes to act as the DAB (as adjudicator) inaccordance with this Dispute Adjudication Agreement.
5. The Engineer and the Contractor jointly and severally undertake to pay the Member, in consideration of the carrying out of these services, in accordance with Clause 6 of the General Conditions of Dispute Adjudication Agreement.
6. This Dispute Adjudication Agreement shall be governed by the law of ____________________

SIGNED by: ______________________________

for and on behalf of the Engineering the presence of

Witness..

Name..

Address..

Date...

SIGNED by: ____________________

for and on behalf of the Contractor in the presence of

Witness____________

Name:____________

Address:____________

Date:______________

SIGNED by: ________________________

The Member in the presence of

Witness:__________________________

Name:_____________

Address:___________

Date:______________

DISPUTE ADJUDICATION AGREEMENT

[for each member of a three-person DAB]

Name and details of Contract

____

Name and address of Engineer

____

Name and address of Contractor

____

Name and address of Member

____

Name and address of Engineer _

____

Name and address of Contractor _

____

Name and address of Member _

____

Whereas the Engineer and the Contractor have entered into the Contract and desire jointly to appoint the Member to act as one of the three persons who are jointly called the "DAB" [*and desire the Member to act as chairman of the DAB*].

The Engineer, Contractor and Member jointly agree as follows:

1. The conditions of this Dispute Adjudication Agreement comprise the "General Conditions of Dispute Adjudication Agreement", which is appended to the General Conditions of the "Conditions of Contract

for Construction" First Edition 2023 published by the Engineer Live Foundation (ELF), and the following provisions. In these provisions, which include amendments and additions to the General Conditions of Dispute Adjudication Agreement, words and expressions shall have the same meanings as are assigned to themin the General Conditions of Dispute Adjudication Agreement.

2. [*Details of amendments to the General Conditions of Dispute Adjudication Agreement, if any. For example:*

In the procedural rules annexed to the General Conditions of Dispute Adjudication Agreement, Rule _ is deleted and replaced by: " ... "]

3. In accordance with Clause 6 of the General Conditions of Dispute Adjudication Agreement, the Member shall be paid as follows:

A retainer fee of________________________________per calendar month, plus, a daily fee of_____________________________per day.

4. In consideration of these fees and other payments to be made by the Engineer and the Contractor in accordance with Clause 6 of the General Conditions of Dispute Adjudication Agreement, the Member undertakes to serve, as described in this Dispute Adjudication Agreement, as one of the three persons who are jointly to act as the DAB.

5. The Engineer and the Contractor jointly and severally undertake to pay the Member, in consideration of the carrying out of these services, in accordance with Clause 6 of the General Conditions of Dispute Adjudication Agreement.

6. This Dispute Adjudication Agreement shall be governed by the law of_____________________

SIGNED by: ________________________________

for and on behalf of the Engineering the presence of

Witness...

Name...

Address..

Date...

SIGNED by: ________________________

for and on behalf of the Contractor in the presence ofSIGNED by: ___

for and on behalf of the Engineering the presence of

Witness:______________________________

Name:_______________

Address:______________________________

Date:________________

Witness______________________________

Name:______________

Address:_____________________________

Date:_______________

www.ingramcontent.com/pod-product-compliance
Lightning Source LLC
LaVergne TN
LVHW031424170726
843492LV00009B/2849

* 9 7 8 8 1 9 6 7 8 4 6 7 6 *